P9-DCM-026

BEFORE THE BIG BANG:

THE ORIGINS OF THE UNIVERSE

BEFORE THE BIG BANG

THE ORIGINS OF THE UNIVERSE

By Ernest J. Sternglass

To see the world in a grain of sand,
And heaven in a wild flower;
Hold infinity in the palm of your hand,
And eternity in an hour.
—*Auguries of Innocence,*
William Blake (1789)

FOUR WALLS EIGHT WINDOWS NEW YORK/LONDON

To Marilyn, who made everything possible

Published in the United States by:
Four Walls Eight Windows
39 West 14th Street, room 503
New York, N.Y., 10011

U.K. offices:
Four Walls Eight Windows/Turnaround
Unit 3, Olympia Trading Estate
Coburg Road, Wood Green
London N22 6TZ, England

First printing October 1997.

Library of Congress Cataloging-in-Publication Data:
Sternglass, Ernest J.
Before the big bang: the origins of the universe/ by Ernest J. Sternglass.
p. cm.
Includes bibliographical references and index.
ISBN 1-56858-087-8
1. Matter—History. 2. Matter—Mathematical models.
3. Cosmology. 4. Particles (Nuclear physics)—History.
5. Physicists—Correspondence. I. Title.
QC171.2.S84 1997
523.1'2—DC21 97-26395 CIP

10 9 8 7 6 5 4 3 2 1

Printed in the United States
Illustrations by Brian J. Lopresti
Text design and composition by Ink, Inc., New York

TABLE OF CONTENTS

INTRODUCTION

WHEN I WAS A YOUNG BOY, two images about the universe made a particularly strong impression on me. At the age of four or five, I was taken to the planetarium in Berlin, the city where I was born. There I learned that a star called Betelgeuse was so huge that it could contain the Sun and the Earth going around it. But looking up into the sky, the stars were just little points, and so I began to realize just how far even the nearest stars must be.

Then, when I was about twelve years old, I saw a cartoon in a popular book explaining basic ideas of physics. It showed the English physicist Sir Ernest Rutherford, who discovered the structure of the hydrogen atom, standing on a ladder and whirling an electron around a proton on an elastic string, demonstrating how the Earth goes around the Sun. I wondered what determined the size of the circle that the electron described around the massive proton: why did this circle have the particular size it did?

These questions about the nature of the building blocks of matter, the sizes of atoms and stars and the distances between them were ones I would struggle with on and off for decades to come. Researching them ultimately led to surprising answers, the essence of which I have tried to explain in the present book.

By the time I was fourteen, just before the Second World War, my family and I had escaped Europe for the United States. After two years of high school in New York, I enrolled as an undergraduate in the School of Electrical Engineering at Cornell University. There I was soon fascinated by the physics of the new vacuum tubes that were being developed to power the still-secret military sonar and radar devices. I joined the Navy for training in this new technology, and when I was discharged in 1946, I accepted an offer to work at the Naval Ordnance Laboratory on the application of electron physics to night-viewing, using the invisible infrared radiation given off by all objects. This work involved studying the way electrons interacted with matter, and in particular how electrons fired at metals and insulators ejected so-called secondary electrons from their atomic orbits.

Within a few months of studying the literature, I began to suspect that the existing theory made certain questionable assumptions about which of the electrons in the atoms participated in the process. But at this point in my life, I only had a bachelor's degree with relatively few formal physics courses, and I needed some reassurance that my ideas about the secondary electron emission process were likely to be correct.

In Washington, where the Navy sent me for final training, I met David Baumgardt, who was Consultant for Philosophy at the Library of Congress and who had been a professor in Berlin, where he had known Einstein. I told Baumgardt of my work and how it was related to Einstein's on the photoelectric effect, in which electrons are knocked out of atoms by light quanta. Baumgardt suggested that I write a letter to Einstein, outlining my ideas on electron emission.

I could not resist this opportunity to meet the most renowned of all living physicists, especially since I had also become interested in the nature of the basic particles of matter and light—an area to which Einstein had made such fundamental contributions. And so, in April of 1947, I wrote a letter about my ideas on electron emission and sent it to Einstein. To my great surprise, I received a reply in less than a week, asking me to come visit him in Princeton.

As described in Chapter Four, it was a visit that had an enormous impact on my whole life. Not only did Einstein's suggestions to me, then and subsequently, support me in my work on electron emission from solids. Perhaps most importantly, they encouraged me to examine the more fundamental problems of nuclear particles. Eventually, this research led me to a new approach to the question of the origin of matter and the structure of the universe. But Einstein's advice to pursue these more difficult ideas in private also had a lasting effect. He advised me to earn a living in some other manner or, as he put it, "to find a cobbler's job" so that I "could make mistakes in private."

This book is an account of the pursuit of a deeper understanding of the nature of matter and its origin in the early universe, together with the many steps forward and backward taken in private as unexpected experimental discoveries and observations were made in the incredible half-century of new particle accelerators, space science and computer technology since my journey to Princeton in April of 1947.

Before the Big Bang is written for the layperson, first and foremost. But I wanted to be sure that people who wished to follow a more detailed version of my argument be able to do so, and to that end I have included a number of precise quantitive predictions. That sometimes involves numbers taken to many decimal places. And to begin with, sub-atomic and macro-cosmic physics are difficult subjects. Nonetheless, I believe my basic concepts will, at the end of this book, be clear to any interested reader.

CHAPTER 1

BEGINNINGS

FOR NEARLY SIX THOUSAND YEARS of recorded history, humans have wondered how the world began, what is the ultimate nature of the matter it is made of, and what will be the fate of the universe. But only within the last few years have we begun to acquire the means for testing speculations about the origin of the universe with definitive observations. Enormous advances have been brought about by the development of space-based observatories and giant new telescopes equipped with novel electronic detectors, revolutionizing astronomical observations.

This book describes a theory of the nature of matter and the origin of the universe. It suggests the cosmos is apparently designed for the development of life and destined to exist forever, neither to fly apart into dying cinders, nor to collapse into a fiery point of infinite density, a "singularity" in which all life will end. If the theory continues to withstand the test of observations and experiments, humanity will continue indefinitely, even beyond the day when the Sun reaches the end of its life, some billions of years from now.

The basic assumption underlying this theory is that the fundamental entities are the electron and its oppositely charged "twin," the positron. These entities form a rotating pair of a very small, but finite size. From one such pair, all other particles and stable cosmological structures such as galaxies and stars arose in a succession of division processes. They were formed out of pure space as vortices in the ether, an all-pervading liquid-like medium first envisioned by the ancient Greek philosophers and revived in the seventeenth century by René Descartes, the founder of modern natural philosophy. The ether concept was used successfully in the nineteenth century to describe electricity, magnetism and light in a historic unification. It was initially rejected by Einstein because all efforts to detect it experimentally had failed. Later, he revived it in his General Theory of Relativity.

According to the electron pair theory of matter, the universe began with an electron and a positron in an extremely heavy, rapidly rotating state, containing the entire mass of the universe in a primordial atom whose electromagnetic field defined physical space.

The idea for a so-called "primeval atom" originated with the Belgian priest and astrophysicist Georges Lemaître in the early 1930s, shortly after the American astronomer Edwin Hubble observed the reddening of light from the majority of galaxies. The reddening suggested that the universe was expanding because of the Doppler Effect, which leads to a lengthening of light waves from a receding source. The lengthening shifts the spectral lines towards the red end of the spectrum. Conversely, the Doppler Effect leads to a shortening of the waves, towards the blue end of the spectrum, for an approaching source.

Assuming the universe was indeed expanding, Lemaître drew the amazing conclusion that at one time the universe had been extremely small, perhaps no larger than an ordinary nuclear particle such as a proton. But the nucleus of a typical atom is ten thousand times smaller than the simplest atom, which is that of hydrogen. The hydrogen atom consists of a negatively charged electron in orbit around a positively charged proton some two thousand times more massive than the electron. Thus, the density of the universe at its moment of initial expansion had to have been truly enormous.

Lemaître's daring idea led to the present "Big Bang" theory of the origin of the universe. But while the prevailing view agrees with Lemaître's general concept, it diverges in holding that the original form of the universe was not a form of matter, but an extremely hot, intense concentration of light or radiation.

At the moment of the Big Bang, the explosive formation of matter composed of protons, neutrons and electrons of ordinary mass is believed to have taken place. After the formation of protons, in the first few seconds, the formation of elements with heavier nuclei began—hydrogen, helium, and lithium.That produced a mass of hot gases, out of which stars and galaxies are believed to have gradually condensed under the action of gravity over many millions of years.

Rejecting the idea of a hot gas as the universe's initial state, Lemaître assumed that the cosmos began with an extremely small but massive particle that divided itself by two in a series of some two hundred sixty steps. Each such split formed a less massive fragment of the original primeval atom from which galaxies and smaller objects such as stars originated in a series of stages. In the last stage of successive divisions, ordinary nuclear particles such as protons and the equally heavy neutron were formed in a cosmic fireworks display.

At the time Lemaître first speculated about the origin of the universe, there was no known form of matter that could take on the required enormous mass of the entire universe in the small volume of a single nuclear particle. He resigned himself to the thought that future developments in nuclear particle physics might some day provide a definite model for such an incredibly dense form of matter. These hopes did not materialize in Lemaître's lifetime, and although many nuclear particles heavier than the proton were discovered during the 1950s, they were increasingly short-lived and not nearly massive enough to provide a model for the seeds of galaxies and stars that he envisioned.

Not until the late 1960s and early 70s were particle accelerators developed that gave very high energies to beams of protons, electrons and positrons. In 1974, a new type of particle was discovered in high energy collisions between electrons and positrons in the two-mile-long nuclear accelerator at Stanford, and also independently found in proton-proton collisions at the Brookhaven National Laboratory. It was given the name J/Psi. J/Psi had a surprisingly long life, thousands of times longer than any of the other particles with a mass greater than the proton discovered up to that time. It was particularly the discovery of J/Psi that caused me to consider seriously an early idea for a possible model of Lemaître's primeval atom: namely, a very compact, rotating, highly excited bound state of an electron and positron pair that could divide into ever-smaller mass pairs with decreasing energy and lifetime. Einstein's famous formula $E = mc^2$ means that high internal energy of motion existing when particles move with a velocity close to that of light *c*, can result in a very large mass *m*. In principle, the Lemaître model was thus able to explain the existence of relatively long-lived, heavy new particles of extremely high density. These particles would be able to form the massive seeds of the known cosmological systems—galaxies and stars.

A theoretical model that I published in 1961 was the first to introduce this concept. A small but massive Bohr atom-like electron-positron system (that is, with an electron orbiting around the positron) could explain the properties of a very short-lived particle discovered in the early 1950s, the neutral pi-meson, which has a mass between that of the electron and the proton. The possibility that this might also turn out to be a model for the Lemaître atom had crossed my mind at the time, but it did not really become believable to me until the day I learned of the J/Psi. I now had a sudden incentive to work out a theory for Lemaître's primeval atom and the process by which he believed galaxies and stars to have originated. It became possible to believe that the universe's beginnings were in a very highly energetic, or "excited" state of an electron and positron moving around each other like a double star, in a very small orbit the size of nuclear particles, with an energy of motion and thus a mass equal to that of the entire universe.

With the hypothesis that the primeval atom was composed of nothing but a massive electron and positron rotating about each other, it appeared possible that during the period of division of the original electron-positron pair that preceded the Big Bang, the ordered structure of the universe was laid down in embryonic form. If this had taken place in a series of stages, each consisting of ten steps of division by two—analogous to the steps between the floors of a tall building—all the way from the seeds of large clusters of galaxies down to stellar clusters, stars and planets, then the observed structure of the whole universe could be laid down in microscopic form within a volume of a grain of sand, as the poet

Blake had prophetically envisioned. Each of these systems would then expand outward starting from a compact "seed" pair that initially contained the entire mass of the system, just as the universe originally did, concentrated in a volume about a trillionth of an inch in diameter.

The idea of a rotating primeval atom dividing by two in a regular manner, giving rise to cosmological systems of decreasing mass in a series of ejection processes, appears to explain some of the most puzzling aspects of recent discoveries made by the Hubble Space Telescope. For example, it seems to explain the pattern of newly forming, compact spherical galaxies arranged in highly regular spiral patterns. Each of these galaxies in turn gradually expands and ejects spiral arms, such as those of our own galaxy, as if they originated from rotating central sources. Moreover, the explosive creation of protons accompanied by other particles in the final stage of division may explain the mysterious gamma ray bursts detected by satellites for decades and recently found to come from the centers of newly evolving distant galaxies, exactly as Lemaître had predicted.

The beauty of this model is that it leads to a universe that will neither fly apart nor collapse after ending its present expansion, as the so-called "standard model" assumes. The electron pair model leads to a stable, finite state of the universe of the form originally postulated by Einstein. The stability of the universe is insured by the fact that the largest system of galaxies, constituting the physical universe, is rotating like all the smaller cosmological systems. This rotation provides a centrifugal force that counters the attraction of gravity. It also affects all the smaller cosmological structures, such as our planetary system, and applies all the way down to the single rotating electron pair which seeded each cosmological structure. This outwardly-directed centrifugal force counters the inwardly-directed force of gravity, which in the case of the universe as a whole explains the mysterious "anti-gravity force" symbolized by the cosmological constant Einstein had to introduce to keep his static, non-rotating model based on his General Theory of Relativity from collapsing.

The idea that the universe may consist of a series of increasingly larger systems, each stabilized by rotation, goes back to the eighteenth century German philosopher Immanuel Kant. But Kant imagined that the number of systems would have no limit, resulting in an infinitely large universe. That the evolution of the universe is continuing seems to be borne out by astronomical observations over the last three decades, which show matter being ejected from compact, massive, active nuclei of galaxies, originally called quasi-stellar objects or "quasars." These extremely powerful objects could arise from as-yet-incompletely divided seed pairs remaining from the Big Bang.

It now seems that new galaxies, producing stars at a very high rate, continue to eject massive fragments of the original Lemaître atom billions of years after the Big Bang. The massive nuclei that have recently been found to exist in the centers of most galaxies, including our own, could be the remnants of the seed pairs from which they formed at the time of the Big Bang. This would account for some of the mysterious "dark matter" that astronomers theorize makes up most of all matter in the universe. These primordial seeds of matter would be expected to survive mainly in the centers of the largest systems, such as superclusters and complexes of superclusters.

The idea that a universe of finite size (as Einstein originally envisioned) could be stabilized by rotation (as Kant envisioned) was revived by the famous mathematician and friend of Einstein, Kurt Gödel, a few years after World War II. However, Einstein did not believe the hypothesis because Gödel had made a mistake that seemed to allow for an unphysical behavior of the equations. This mistake in Gödel's calculations was only detected and corrected in 1962 by the astrophysicists Ozsvath and Schücking, seven years after Einstein's death.

The electron pair model leads to a stable proton with an internal "quark" structure ultimately composed of nothing but relativistically moving (and thus massive) electrons and positrons that circulate in a highly synchronized manner and seemingly produce permanent stability. Such a relationship between electrons and quarks was stipulated by the so-called Grand Unified Theory developed in the 1970s. As recently explained by the British theorist Stephen Hawking in *A Brief History of Time*, at sufficiently high energies "matter particles like quarks and electrons would all be essentially the same." This is consistent with a type of quark suggested by the Japanese-American theorist Voichiro Nambu in 1965. Nambu's quark has the same charge as the electron, instead of carrying a charge of one-third or two-thirds of an electron, as assumed by Murray Gell-Mann in the original form of his theory of the strongly interacting nuclear particles.

Such a "sub-structure" for the heavy quarks, which in the electron pair model would consist of heavy electrons and positrons, seems to be supported by the recent results of high energy collisions between positrons and protons as well as protons and anti-protons, the negatively charged twins of protons. And an electron pair structure of quarks, and thus the heavy nuclear particles, explains both why no "fractional" charges smaller than that of the electron have ever been found despite thirty years of intensive search, and why the proton has exactly the charge of the positron to less than one part in a billion.

An electron-positron structure for the proton is also strongly suggested by the fact that, whenever protons are seen to annihilate with oppositely charged anti-protons in the laboratory, the short-lived mesons that are produced always decay into electrons and positrons, together with various forms of radiation, but never into fractional charges.

The great stability of the proton is supported by more than a decade of experiments that have failed to find a spontaneous decay of protons, or any evidence for its disintegration. This stability means that there is no limit to the length of time that matter in its present form can continue to exist. The electron pair model therefore allows an unlimited future for conscious life as we know it, in a closed universe where energy is conserved and the collisions of galaxies give birth to new stars and planets.

An equally attractive feature of a theory based on the electron and positron as the basic entities of all matter is that no arbitrary or adjustable quantities are needed to explain the mass of the universe, its size, and its highly regular organization other than the fundamental atomic constants that specify the electron and its antiparticle. Thus, in sharp contrast to the "standard model" that requires more than a dozen experimentally-determined parameters, the only quantities essential to the electron pair theory are the basic atomic constants: (1) the mass of the electron, (2) the charge of the electron, (3) the maximum speed of the electron (equal to that of light), and (4) the spin of the electron given by Planck's Constant, a quantity that is at the root of quantum theory.

Given just these four fundamental constants that appear not to have varied throughout the evolution of the universe, the masses of all the known stable and unstable nuclear particles can be derived as energized states of the proton, or as molecular-like systems of relativistic electron-positron pairs. These four fundamental constants of atomic theory also suffice to specify the average masses, sizes and spacing of all astronomical objects in their surprisingly ordered arrangement in the heavens, as well as the mass of the universe and the value of Isaac Newton's gravitational constant within the limits of present observations.

The theory combines the fundamental features of classical mechanics as derived by Galileo, Kepler, Descartes and Newton with the basic aspects of electrodynamics, Einstein's relativity and quantum theory. It does not demand discarding detailed space-time geometric models of matter as Niels Bohr, Werner Heisenberg and Max Born advocated. Einstein, who stubbornly insisted on a "causal," detailed geometric description of atomic phenomena, in opposition to Bohr and many other physicists, may thus be vindicated.

It now appears that the electron, the positron and the quanta of light from which they can be formed can be described in terms of nothing but stable vortex rings in the ether, a fluid-like medium that constitutes space. Like smoke rings or a tornado, the uniform density of the vortex rings is distorted by their internal circular motion. Indeed, a vortex ring is nothing other than the funnel of a tornado bent to form a closed tube. As a result, the physical origin of "space curvature" associated with matter in Einstein's General Theory of Relativity is explained by the internal circulation of a fluid in a vortex ring.

Similarly, the spin of an entire vortex ring (or an electron-pair) distorts space. Every form of motion in the ether causes it to become deformed, creating "dimples" in ordinary "flat" or Euclidean space, just as a spoon gently drawn across coffee in a cup forms a vortex. In 1858, Helmholtz used this analogy at the end of his groundbreaking paper on the stability of vortices in an ideal fluid. But of course he did not mention the obvious—that you can also create a disturbance in the cup by simply stirring it.

With matter in the form of electrons and positrons, and photons regarded as nothing but forms of quantized, stable rotational motion in an ideal, frictionless fluid, it becomes possible to understand how particles of matter and light necessarily lead to a curvature of space around them that Einstein showed to be equivalent to the action of a gravitational force.

Ironically, it was the rejection of such a universal fluid or ether by the young Einstein (which he subsequently regretted) that made it impossible to accept the idea of matter and light as forms of stable vortex rings. Yet nineteenth century giants of science such as Helmholtz, Maxwell, Kelvin and Thomson believed in these vortices that move like point particles, but are actually entities that extend like waves through physical space. Such a geometric model of matter and light explains one of the greatest mysteries of quantum theory, the wave-particle duality. This is the puzzle presented by the particle nature of light found by Einstein in 1905 and the wave nature of particles arrived at by de Broglie in 1923. The vortices can be identified with the superstrings that are equivalent to stable tubes of vortex motion of finite diameter, the basic geometrical entities from which all particles are made. One of the superstring theorists, Edward Witten, has said that "the great ideas in physics have geometric foundations."

As Timothy Ferris points out in his recent comprehensive account of modern particle physics and cosmology, "all 'forces' are consequences of geometry." It is a prospect that was raised by Bernard Riemann in the middle of the ninenteenth century, by Einstein to explain gravity, and more recently extended by superstring theory to the three other forces, the electromagnetic, the strong and the weak nuclear force. As Ferris puts it, superstring theories are in effect "unified" theories, combining general relativity and quantum physics.

Until now, the diameters of superstrings have been found to be far too small to have any relation to ordinary nuclear particles or photons, being smaller than the proton by some twenty-one orders of magnitude, or a billion trillion times too small. But in the electron pair model, when the local gravitational curvature constant grows with the decreasing mass of the electron pairs as they divide again and again, the diameters of the superstrings grow to those of electrons and the other known nuclear particles. At this point, the strength of the local gravitational force grows to equal that of the electromagnetic force. Thus it becomes strong enough to stabilize a finite-sized electron against the repulsive force due to its charges, as first suggested by Einstein in 1919 and worked out in detail by the astrophysicist Lloyd Motz in 1962.

Interpreted in terms of the properties of quantized vortices that are identified with the electromagnetic field between electrons and positrons, superstrings do in fact have many beautiful properties. A theory of matter regarded as strings or tubes of quantized vortex motion draws all matter into a single elegant picture. As Ferris puts it so well, in such a theory "particles' attributes are seen as the vibrations of strings, like notes struck on Pythagoras' lyre." It is a concept that applies exactly to the field-lines of electron pair systems rotating near the speed of light that can be excited to vibrations that have an energy reaching from those of nuclear particles to those of planets, stars and galaxies. They thus represent the modern equivalent of the harmonies of the Pythagorean spheres.

Mathematically, finite-sized sources of these lines or tubes of force do away with the infinities that dogged quantum field theory, where electrons were regarded as dimensionless points. As Ferris notes, superstrings of finite diameters dispense with "the dirty business of renormalization," a mathematical procedure to remove the infinities in quantum field theory developed in the 1930s and 1940s.

The electron pair model of matter provides a natural explanation for the occurrence of ten "dimensions" of space-time, six more than the three dimensions of our ordinary physical space and the one dimension of time of Einstein's relativity theory. One of the additional dimensions is that associated with the small "curled up" but finite size of the space defined by the existence of an electromagnetic charge, an idea going back to those of Theodor Kaluza and Oskar Klein in the 1920s. In the electron pair model, the size of the inner space is equal to the diameter of the vortex tube from which a pair is formed when a closed vortex ring or superstring is cut in two. This happens when opposite sides of a vortex ring come together and overlap, either as a result of spontaneous vibrations or an energetic collision process. The remaining five "dimensions" can be identified with the so-called "degrees of freedom" associated with the three sets of rotational motions of the electron pair systems forming the basic structure of protons and neutrons, together with the two degrees of freedom associated with the highly relativistically moving electron and positron that stabilize the proton and bind it to the neutron, all located within the "curled-up" dimension defined by the size of the electron.

This model of matter also explains the inflation undergone by the universe in its earliest stages before the Big Bang, as first suggested by Alan Guth in 1980. In terms of the present model for the initial state of the universe, the process of repeated division of pair-systems is similar to that of cells in a living organism. In such a division by two, the number of electron pairs separated by a minimum distance grows exponentially. Thus the accelerating expansion of space was driven by the division process, in which the pairs got rid of their enormous energy density by dividing into two pairs, each of which had half the original energy or mass. This process of increasingly rapid expansion ended after two hundred and seventy divisions, just ten more than Lemaître had estimated, at which stage the pairs had each decreased in mass to about twice what is needed to form two protons.

At this point, there was a change in the arrangement of the pairs that led to an extremely stable, crystalline-like state that results in protons, a process called a phase transition, analogous to the transformation of water molecules into crystals of ice. In this geometrical rearrangement, four sets of pairs were locked into a state where a relativistically-moving and thus massive positron is exchanged between two such pairs as a unit, producing an extremely stable proton structure composed of three parts, named "quarks" by Murray Gell-Mann in 1964. Without this change in the arrangement of the pairs into stable quarks and protons, the universe would have ended in the mutual annihilation of the electrons and positrons into pure radiation in the form of gamma rays, a process observed in the laboratory shortly after the discovery of the positron in 1931. It is the reverse of the production of a pair of charges from a single photon or gamma ray also discovered in the early thirties, a process that can be visualized as a closed vortex ring colliding with a proton so that the sides of the ring are forced together, cutting the ring and thus creating a pair of charges.

In the process of the formation of stable protons, neutrons and electrons of ordinary mass, the energy not used in the formation of these particles became available as motional energy given to the particles. This process heated the early universe to something like a trillion degrees centigrade to form the hot Big Bang. This was the derisive name for the explosive creation of ordinary matter given to it by the British physicist Fred Hoyle. Hoyle did not believe in Lemaître's idea of a single creation process, and had proposed a "steady-state theory" in which the universe had no beginning or end and did not change in time, with new matter being constantly created throughout physical space to keep the universe unchanged in density despite its expansion.

The first few minutes after the Big Bang, after the lightest elements were formed, the universe cooled too much to sustain the formation of the heavier elements, which were subsequently created in the stars. According to the electron pair theory, after the first second, the universe evolved pretty much per the standard Big Bang model as described by theoretical physicist Steven Weinberg in *The First Three Minutes*—except for one significant difference. Not all the electron pairs that formed from the first pair divided immediately all the way to the final stage, where protons and ordinary low-mass electrons came into being to create a uniform hot gas consisting mainly of hydrogen and helium. Some of the massive electron pairs remained trapped in clusters, initially containing about a thousand pairs in a volume roughly the size of an ordinary atomic nucleus, or about a trillionth of an inch in diameter. These provided the massive seeds around which the various cosmological systems were produced by the gravitational infall of the hydrogen and helium gases formed by the fraction of the massive pairs that had escaped from their central clusters. Such seeds, resulting in a small non-uniformity in the distribution of matter in the very early universe, allowed the collapse of the newly created gases to form galaxies and stars in the relatively short time of a few billion years.

In fact, as described by Guth in his recent book *The Inflationary Universe*, the discovery of a small non-uniformity in cosmic background radiation has been one of the major pieces of clear-cut evidence for the Big Bang. But there has been a problem accounting for the small irregularities in the radio waves that represent the cooled radiation from the Big Bang. When it is assumed that this non-uniformity is due to some sort of random quantum fluctuations in the early universe, it becomes impossible to explain the surprisingly regular arrangement of galaxies and clusters of galaxies that have been discovered in the last few years. But the regular division process in twenty-seven major stages of ten divisions by two in the electron pair model can explain that geometrical relationship. Nor has there been any success in detecting new kinds of "cold dark matter particles" that are assumed to have provided the centers around which galaxies and stars were formed in the early universe. However, the massive electron pairs that resulted from the regular division of the primeval atom explain the nature of the heavy dark matter particles needed to form the seeds of these cosmological systems. In fact, among the lowest mass examples of such dark particles that do not decay with the emission of light are the J/Psi and Upsilon mesons. Produced in electron-positron collisions in high energy accelerators, these mesons, like the primeval atom, decay by dividing, and thereby forming less massive mesons that are composed of nothing but rapidly rotating, massive electrons and positrons.

All scenarios that begin with a period of accelerated expansion preceding the formation of protons in the Big Bang remove certain conceptual difficulties that the early Big Bang model encountered. These include the prediction of particles called "magnetic monopoles" that have never been observed, and the difficulty of understanding the high degree of uniformity of the universe on a sufficiently large scale in all directions. Inflationary models that start slowly from a very small region of space and reach a size much larger than the presently visible universe can also overcome the problem that space appears to be nearly Euclidean or "flat" out to the largest distances observed so far, despite the expectation derived from general relativity that it should be sufficiently curved so as to be closed. A nearly Euclidean geometry also implies a very critical, extremely "fine-tuned" initial expansion energy that leads to exactly the critical density of matter that causes the universe neither to fly apart to an infinite size nor to collapse again into a point singularity. This instability does not exist in a universe that starts as a rotating electron pair, where the attractive forces are balanced by centrifugal forces no matter how high the internal energy is, and thus ends up with all its hierarchically-organized systems in rotational equilibrium regardless of their density.

Another important difference between the inflationary and the electron pair model is that in the latter there is no infinitely dense state or singularity at its beginning. Such a singularity was shown by Hawking and Roger Penrose to be inevitable if one uses only the classical General Theory of Relativity, in which quantum phenomena have no place. But when quantum theory comes into play, as it does in the actual physical universe through a maximum possible density for an electron pair of finite size, this singularity disappears. This is the Planck Density, of enormous but finite value, equal in grams per cubic centimeter to the number five followed by ninety-three zeros, as first suggested by the Russian theoretical physicist V. L. Ginzburg in 1971. At this density, space is believed to tear and develop holes, creating charges.

In the electron pair model, at this density a closed vortex ring or superstring vibrates so violently that it pinches and cuts itself in half. The two ends of the halves rotate one hundred eighty degrees relative to each other and cling together, due to their low internal pressure, forming the first pair of charges—or the primeval atom that Lemaître envisioned more than sixty years ago. The volume of the region into which the field-energy associated with the electron pair is concentrated, together with the Planck Density, gives a precise theoretical value for the mass of the universe that is consistent with the value obtained from an estimate of its present density and size. Moreover, using the density of the universe given by the model, one can calculate the value of the Hubble Constant that gives the present rate of expansion and thus the age of the universe since the Big Bang, in good agreement with the most recent observations.

The process of division of the first electron pair by two in stages of ten such steps on average leads to a highly regular pattern in the masses of cosmological systems. Every division process produces two such pairs, each with a mass half as large. After ten such divisions there will be 1,024 pairs, each with a mass 1,024 times smaller than the original pair. Thus, the seed-pair for a supercluster will result in about a thousand galaxies making up such a system, each with a mass about a thousand times smaller than that of a supercluster. Similarly, the seed-pair of a galaxy such as ours will eventually give rise to a thousand dwarf galaxies. Each seed-pair of a dwarf galaxy will in turn give rise to a thousand globular clusters, and so on down to stellar clusters of about a thousand stars. Moreover, the model requires that the sizes or spacings of these large systems will decrease by a factor equal to the square root of the factor by which the masses decline in ten divisions, or a factor of thirty-two.

Such a relation between the masses and sizes of objects has in fact long been suspected. It supports the theory of Descartes, that the stars in heaven are aranged in systems of vortices in a fluid-like ether.

The patterns of galaxies and stars strongly support a vortex or superstring model of matter, and an origin by successive division from a single primordial electron pair of finite mass and size. In this model, there was no infinitely dense singularity at the beginning of time, and matter, like light, consists of nothing but localized motional energy.

The electron pair model allows one to calculate the periods of rotation of the various massive seed pairs, and thus the time it takes for them to divide as their masses decline in the course of the division process. This leads to the surprising conclusion that the most massive of all pairs, Lemaître's primeval atom with a mass equal to that of the entire universe, corresponding to that of some seven trillion galaxies and a radius of 2.5 trillion light-years—hundreds of times larger than the present visible radius of about ten billion light-years—requires a period of some fifteen trillion years for a single rotation. Instead of a small fraction of a second in which Guth's inflation is assumed to have taken place, it now appears that some fifteen trillion years passed before the original seed of the universe divided and gave rise to ordinary matter in the accelerating expansion that ended with the Big Bang. Although the birth of ordinary matter as we know it took place some ten billion years ago in a state of enormous but finite density, it seems that the formation of ordinary matter was preceded by a long period of gestation, during which the structure of the universe was laid down in embryonic form.

When there is no singularity in a theory for the origin of the universe, there is no point where the laws of nature break down. It is therefore possible to believe that the laws of nature were laid down by a Creator before the first pair came into being. As Hawking puts it: "Science seems to have uncovered a set of laws that, within the limits set by the uncertainty principle, tell us how the universe will develop with time, if we know its state at any one time." Hawking also concludes that if the hot Big Bang model is indeed correct right back to the beginning of time, "it would be very difficult to explain why the universe should have begun just in this way, except as the act of a God who intended to create beings like us."

The electron-positron pair model of matter now seems to provide us with just such knowledge of the state of the universe at the beginning of time. We may be approaching a stage when we can begin to answer the question that Einstein regarded as the most important one: namely, whether God had any choice in establishing the laws of nature.

The result is that the universe, far from being a random or chaotic system without order, structure and coherence, the basic entities of which cannot be visualized, appears instead to be amazingly comprehensible and rational, and far more like a living, evolving entity that is governed by surprisingly simple laws. It appears to be designed according to a simple architectural principle capable of evolving complex structures and self-replicating organisms based on the inherent properties of its fundamental constituents in accordance with unchanging laws obeyed on every scale. In its development, the universe appears to be a still-evolving entity that is governed by surprisingly simple mathematical relations understandable in terms of fluid, geometric space-time models without being locked into a machine-like pattern of rigid predetermination. The story of the detailed discoveries and theoretical developments that have opened up this possibility in the last one hundred years form the subject of the present book.

CHAPTER 2

FARADAY'S DREAM

WITHIN THE LAST THIRTY YEARS, the evidence has mounted that the universe is in a state of explosive expansion, with galaxies racing away from each other exactly as Hubble concluded in the late 1920s. It follows that at one time in the distant past all matter must have been concentrated in an incredibly small volume of enormous mass. But just how small the volume was is the question that remained a mystery, closely tied up with the mystery of what the ultimate constituents of matter are. As increasingly powerful particle accelerators have produced an ever-growing number of new and unexpected particles, questions have only proliferated. As a result, both fundamental particle theory and cosmology are in a state of crisis which is typified by the problem of the apparent existence of an invisible form of matter, the so-called "dark matter" that seems to make up most of the mass of the universe but whose relation to ordinary matter is not understood. Thus, the current crisis cannot be resolved until the question of the nature of the ultimate particles of matter has been answered.

The idea that there must be some sort of ultimate or indivisible particles from which all ordinary objects are composed is very ancient, going back to the Ionian philosophers who lived some 2,500 years ago along the Mediterranean coast of Asia Minor. At the turn of the present century, the experiments of the English physicists Joseph J. Thomson and his student Ernest Rutherford had led to the conclusion that the basic, indivisible entities which composed all matter (as proposed by the Ionian philosophers) were the negatively charged electron and the positively charged proton, which has a mass nearly two thousand times that of the electron. Both particles appeared to be stable and to behave like indivisible little billiard balls of some "ponderable" mass charged with electricity. The only difference seemed to be their opposite but equal electrical charge and their different masses.

In 1913, with the Danish physicist Niels Bohr's successful theory for the structure of the simplest atom, that of hydrogen, as an electron orbiting around the massive proton, it seemed that the problem of the ultimate constituents of all matter had been solved. Just as the Earth circulating around the Sun is held in its orbit by the equilibrium between the attractive force of gravity and the opposing centrifugal force resulting from the orbital motion—as a stone is whirled around on a string—the electron moves in a circular orbit, but with the tension of electric lines of force holding it in place.

Bohr had gone beyond classical Newtonian theory by assuming that the electron could move only with a certain velocity and angular momentum in its normal, stable orbit as defined by Max Planck and Albert Einstein's quantum theory. But Bohr's model for the hydrogen atom was essentially a simple, classical, mechanical one. And since, in the early 1920s, the same basic model was found to explain all the heavier atoms making up the different chemical elements in the next decade, it seemed possible to explain all forms of matter in a very simple, unified manner in terms of electric interactions alone. However, the difference in mass between these two basic particles and its physical origin remained a puzzle.

Scientists could simply accept the idea of some sort of unknown measurable matter in the form of hard spheres as a fact of nature, with different amounts of matter in the electron and proton but with equal and opposite electric charges "painted" on their surface. This would allow physicists to describe the hydrogen atom and all other more complex atoms very successfully, as was in fact done by Bohr .

On the other hand, it was possible to regard the electron and proton as nothing but very small sources of electrostatic fields of different sizes, without any "hard" or "ponderable" matter, as first suggested by the Slavic scientist Boscovitch in the eighteenth century. This was an idea favored by Michael Faraday, the renowned English scientist whose experiments laid the foundation for the unification of electricity, magnetism and light in the mid-nineteenth century. Faraday developed the idea of lines of magnetic force that originate from one pole of a magnet and terminate on the other, opposite pole, and electric lines of force that go from a negatively charged metal sphere to a positively charged one. The strength, or density, of these lines indicates the direction of the force.

In the early 1890s, the energy that it took to establish an electric field in the space surrounding a spherical charge was shown by the Dutch theoretical physicist Hendrik Lorentz to make it possess inertia or resistance to a change in motion, a property we associate with mass. Thus, mass might be only a manifestation of energy in the region around a source of an electric field or "charge," with nothing but empty space in the center.

A hurricane or tornado, which carries an enormous amount of energy in a rapidly rotating column or vortex of air, demonstrates such a theory of the nature of mass. Ordinarily, when in a state of rest, air offers little resistance to motion. But when it circulates at very high velocity to form a vortex, it can become a tornado. The funnel of a tornado moves along at very high speed and acts as if it is a large mass. It can lift houses, flatten forests and towns. At the center is the non-rotating, quiet eye. Nevertheless, because the energy in the localized circular vortex is so huge and the entire rotating funnel moves as a unit, it acts as if it were a solid object of great mass and momentum. Its momentum is a measure of the mass times the velocity of the object and thus is related to the total energy carried by the tornado.

It is possible to identify the source of the electric field with a stable, microscopic tornado, which carries momentum and has a certain mass associated with it. Its origin lies in the motion of a hypothetical fluid medium that itself has no mass. The resistance to any change in the linear motion of this microscopic vortex measures the mass or energy it contains. For the proton, the size of the source or the eye of the tornado would simply be a few thousand times smaller or more compact than for the electron in this simple model. However, as will be discussed later, this explanation for the greater mass of the proton failed to stand up to experiments carried out beginning in the 1950s. But the model has held up for the electron, and in the early part of the century such a fluid, hydrodynamic or purely electrodynamic view of all matter appeared to be a real possibility.

In the mid-nineteenth century, the Scottish physicist James Clerk Maxwell developed the mathematical theory of electromagnetism based on Faraday's idea of electric and magnetic fields. Together with Einstein's work on gravity—which he had explained in terms of a distortion of space in his General Theory of Relativity in 1915—many scientists began to regard the electron and proton as nothing but stable sources of electric fields. The masses of these particles could be regarded as simply a manifestation of the energy stored in their fields, analogous to the motional energy in a vortex. This idea of a purely electrical origin of mass as a form of localized energy also accorded with the idea of the equivalence of mass and energy developed by Einstein some ten years earlier. It appeared possible that mass was just a manifestation of the energy in the electric field around the two charges, the only difference being the sizes of the source of the field. At least in the case of atomic phenomena where gravity appeared to play no role, the theory of electrical and magnetic interactions between particles as developed by Maxwell and others appeared to be all that was needed to describe the properties of matter.

The first difficulties for such a simple, purely electrical model for matter arose in the early 1920s, when Rutherford and his students were trying to explain the structure of the massive nuclei of the atoms heavier than that of hydrogen, which had only a proton making up the central core. They concluded that in order to explain why the masses of all the heavier nuclei were greater than could be accounted for by the number of positive charges or protons, typically twice as great, there must also be neutral particles in the nuclei of the atoms heavier than hydrogen whose masses were close to that of the protons.

One possibility was that such a neutral particle or neutron, as Rutherford named this hypothetical particle, was composed of an electron and a proton in a new and as yet not understood form, perhaps in a small orbit some one hundred thousand times smaller than the orbit of the electron in Bohr's planetary model for hydrogen.

But even if electrons were to exist in the nucleus of the atoms so as to make neutrons out of protons, there were other theoretical difficulties that seemed impossible to explain in terms of ordinary electromagnetic forces acting by themselves inside the heavier nuclei. Calculations showed that the repulsion between the remaining positively charged protons would be so strong, that unless the protons interacted with the neutrons and each other with a much stronger new kind of attractive force, the nuclei would be unstable and break up. Thus, it seemed that the structure and stability of the nuclei heavier than hydrogen required the existence of a force other than the electromagnetic one, the so-called strong nuclear force, and the early hope for a simple, unified description of matter in purely electromagnetic terms would have to be abandoned.

In 1932, this conclusion was greatly strengthened. After nearly a decade of fruitless search, one of Rutherford's students, James Chadwick, carried out an experiment that proved that the neutron actually existed. Early that year he had learned of an experiment carried out by Jean Frederic Joliot-Curie and his wife Irène Curie, the daughter of Marie Curie, who had discovered radium in 1898. The French scientists bombarded a beryllium nucleus with powerful alpha particles (the name given to particles made of two neutrons and two protons), emitted by radium. Emissions from the beryllium in turn knocked out protons from a paraffin wax target. They concluded that an extremely powerful gamma ray similar to an X-ray or photon was sent out by the beryllium nucleus.

In repeating the experiment, Chadwick found that the results could be explained much better if a neutron—instead of a gamma ray—had been ejected from the beryllium nucleus, which then knocked out a proton from the paraffin. But the neutron could not be regarded as a simple composite entity made up of a proton and an electron, a theoretical difficulty that had to do with the neutron's spin, analogous to that of an ordinary spinning top. That conclusion seemed to deny an electromagnetic explanation of matter.

Almost ten years earlier, two Dutch scientists, George Uhlenbeck and Samuel Goudsmit, had concluded that the electron had to be spinning with an angular momentum equal to half the orbital angular momentum of the electron in Bohr's model for the hydrogen atom. This reduced spin was required to explain the detailed aspects of the light emitted when the electron was excited to more energetic states. Moreover, new measuring techniques developed during the 1920s showed that the proton had the same spin as the electron, and that the neutrons in the atomic nuclei acted as if they also possessed the same "quantized" or fixed, universal half-unit of angular momentum.

Uhlenbeck and Goudsmit had found that in order to explain the pattern of spectral lines, the spins of particles could only be oriented either in the same or in the opposite sense. Similarly, two pieces of magnetized steel will line up spontaneously, either with their north-poles pointing in the same direction when arranged in a single line, or pointing in opposite directions when arranged next to each other. This meant that if the neutron was a composite entity as Rutherford had originally assumed, it would have to have a net spin of either zero or one, depending on the orientation of the two spin angular momenta.

However, the measurements of the spins of the atomic nuclei of elements containing many neutrons and protons indicated that the neutron did not have a spin 0 or 1, but that it had to have a spin of one-half. There seemed to be no way that an electron of spin one-half and a proton of spin one-half could form a neutron with the required spin one-half.

Thus it seemed certain that a promising, simple, unified view of matter consisting of just the electron and the proton interacting with purely electromagnetic forces had to be given up. The discovery of the neutron with a spin of one-half appeared to end the hope that matter could be understood in purely electromagnetic terms as Faraday, Maxwell and Lorentz had hoped.

But there were more reasons why it looked as if electromagnetic theory—even as modified by the discovery of quantum theory—could not explain the newly discovered neutron and nuclear phenomena. In investigating the nature of radioactivity discovered by Becquerel in 1896, where certain chemicals ejected powerful particles, scientists gradually realized that there was a serious problem with the conservation of energy.

After Rutherford and Bohr had developed the planetary model of the atom with massive small cores or nuclei in the center surrounded by orbiting electrons, it was evident that these very powerful forms of radiation had their origin in the central nuclei and not in the outer shell of orbiting electrons. The problem of the emission of alpha rays from the central nucleus of heavy atoms was solved by the Russian-American physicist George Gamow with the help of quantum theory. And gamma rays came to be understood as quanta of light or photons radiated by the protons in unstable states of complex nuclei, much as the less energetic photons were produced by the vibrations of orbital electrons in the outer shell of excited atoms. The difficulty was connected with the emission of the beta rays, electrons shot out of certain unstable nuclei at energies as high as a million volts, which were thereby transformed from one chemical species to another with a greater positive nuclear charge. It turned out that after highly precise measurements of the mass of atoms and the energy carried away by beta particles had been perfected during the 1920s, there appeared to be cases where there was missing mass or energy following the ejection of a beta ray. The sum of the mass-energy carried away by the beta rays and the mass-energy of the resulting atom did not add up to the mass-energy of the original atom that had emitted the beta particle.

At first, it seemed as if the long-established principle of the conservation of energy would have to be abandoned in the case of the emission of electrons from the nuclei of unstable forms of atoms. But as proven again and again during the previous century, this principle seemed so useful in understanding all ordinary chemical and heat producing processes that even though some physicists such as Niels Bohr were willing to consider abandoning it, others such as the Austrian theoretician Wolfgang Pauli refused to give it up. So, in order to account for the missing mass-energy in the emission of electrons from nuclei in the process called radioactive beta-decay, in 1931 Pauli made the daring assumption that there must be a new form of non-electromagnetic particle or quantum of radiation emitted together with the electron. This particle, dubbed the neutrino or "little neutron" by the Italian-born physicist Enrico Fermi the following year, was assumed to carry away the missing mass-energy. The neutrino passes through even the heaviest walls of detecting devices, so that it was not possible to prove its existence directly for more than two more decades after Pauli's proposal.

At the time, many scientists thought this was a desperate attempt to save the principle of the conservation of energy. Pauli was postulating a hypothetical new kind of entity that seemed to be totally unrelated to any other known particles of matter or electromagnetic radiation, an entity invested with properties that appeared to guarantee that it could never be observed directly in the laboratory. But giving up the conservation of energy was such a major break with all the well-established, well-tested laws of physics that Pauli's bold suggestion was taken seriously by many scientists. In particular, Fermi decided to try to work out a detailed theory for the emission of neutrinos by analogy to the emission of photons. In his theory, he assigned a different, much weaker strength to the source of the waves than that of normal electrical charges, which oscillate back and forth to emit ordinary electromagnetic radiation.

Fermi's theory, published two years after Pauli's suggestion, turned out to be so successful in explaining the emission of beta rays accompanied by neutrinos from many different nuclei that the existence of yet another non-electromagnetic force (besides the strong nuclear force) acting within nuclei was difficult to deny. Thus, by the mid-1930s, there appeared to be three unrelated forces in nature besides gravity, making a total of four ways in which particles interacted: the electromagnetic force associated with the electron, the strong nuclear force acting between protons and neutrons, the weak nuclear force associated with the neutrino, and the weakest of all four forces, that of gravity, believed to play no role in the nucleus or in the interaction between electrons.

The newly discovered neutron spontaneously decayed into a proton and an electron with a loss of momentum and energy as well as spin—in a process that could only be described if one accepted the idea of a neutrino emitted simultaneously, in accordance with Fermi's theory.

There were other reasons, Einstein notwithstanding, why it was seemingly hopeless to pursue Faraday's dream of a unified theory of matter based only on electromagnetism and gravity. To begin with, on the basis of classical electromagnetic theory alone, it was difficult to understand how electrons could be fitted into the small nuclei that were not much bigger than the source of a purely electromagnetic electron. As the analogy to a tornado suggests, even an electron regarded as nothing but a stable source of an electromagnetic field has to have a finite size, as does the funnel of a twister and its surrounding moving air that contains its energy and angular momentum. The electron's size turns out to be comparable to that of a small nucleus containing a few neutrons and protons. In fact, immediately after the discovery of the electron in the mid 1890s, it was realized that an infinitely small charge or source of an electric field would have an infinitely large mass associated with it, a problem that has haunted theoretical physics ever since.

But the development of quantum theory added yet another difficulty to the problem of how to squeeze an electron into a neutron or a nucleus. In 1905, Einstein had postulated that light consisted of discrete particle-like quanta or bundles of electromagnetic energy of a finite size that decreased as their energy increased. In 1923, the French physicist Louis de Broglie theorized that electrons would also have to be regarded as a wave-like pulse that became smaller as the electron was made to move faster.

In effect, this compounded the problem of how to fit electrons into a neutron or nucleus. The reason was that whenever electrons or beta rays were ejected from a neutron or a nucleus of an atom, they never emerged with an energy of more than a million volts or so. As great an energy as this was, it was not nearly sufficient to shorten the electron wave enough to fit into an orbit the size of a neutron.

In fact, it was easy to show that it would take energies on the order of a hundred million volts or so for the de Broglie wave associated with an electron to be forced into a region that small. The observed beta-decays shot electrons out of the nucleus with less than one hundredth of what was required, according to the wave theory of de Broglie. Ironically, only a few years later, the wave theory of the electron was confirmed in experiments carried out by George P. Thomson, the son of J. J. Thomson who had discovered it, and independently by two American scientists, Clinton Davison and Lester Germer. By the end of the 1920s, it appeared that Rutherford's idea of a model for neutrons being composed of electrons and protons held together by electromagnetic forces in a new kind of nuclear orbit had to be given up.

Accordingly, it seemed to most physicists that the actual experimental discovery of Rutherford's hypothetical neutron, coupled with the need for still another neutral particle, the neutrino, each with a separate new non-electromagnetic force, was the final blow to the attractive idea that mass was nothing but a manifestation of localized energy in the electromagnetic field surrounding a charge. Einstein's long efforts to arrive at a unified theory in which electromagnetism was the only force other than gravity appeared doomed to failure.

This was a time when such classical, geometric, mechanical or dynamic models for explaining phenomena on the atomic or nuclear scale as Einstein believed in were not in vogue. So circumstances were ripe for a totally different, purely formal approach to the problem of the neutron and proton. Such an approach was developed by the German theoretical physicist Werner Heisenberg.

By 1925, Heisenberg had developed a new kind of abstract mathematical quantum mechanical approach to the description of atomic phenomena that did not rely upon visualizable models. It was called "matrix mechanics," and used arrays of numbers in regular rows and columns to predict the intensities of the different light-waves emitted from atoms as excited atoms jumped from one allowed orbit to another. At the same time, the Austrian scientist Erwin Schrödinger had begun to apply de Broglie's idea of the wave nature of the electron to explain the properties of the hydrogen atom. Schrödinger used familiar partial diffential equations from many areas of physics—such as ones applying to the vibration of strings or drums—in a classical, visualizable approach to quantum theory that became known as wave mechanics.

Heisenberg's abstract approach, developed further by Max Born, Pascual Jordan and Wolfgang Pauli, was later shown to be equivalent to Schrödinger's wave mechanics as based on de Broglie's theory. However, it had proved difficult for Bohr and others to work out detailed, visualizable orbital models for atoms more complex than hydrogen. Heisenberg, who had been working as a young assistant to Born, persuaded Bohr of the power and elegance of the abstract algebraic rather than the visual, geometric approach. The Bohr version of quantum mechanics became widely accepted as the way to tackle the puzzling new atomic and nuclear phenomena.

A few years later, this matrix approach to atomic phenomena was bolstered by the success of the British theoretician Paul A. M. Dirac. Dirac developed a form of the wave equation that incorporated Einstein's special theory of relativity into the description of the spectral lines produced by atoms and at the same time showed that the spin of the electron appeared automatically in the process.

A further boost to the purely formal, abstract descriptions of phenomena on the atomic and nuclear scale was provided by other developments during the historic decade of the 1920s. One that had a particularly great impact on physical and philosophical ideas was Heisenberg's Uncertainty Principle, published in 1927. This principle was based on the existence of a new universal constant related to angular momentum or momentum times distance called "action," which Max Planck first postulated to explain the existence of "discrete" (or discontinuous) steps in the emission of light from atoms and molecules.

This idea of a discontinous emission of light was extended by Einstein in 1905. He suggested not only the existence of distinct steps of emission or absorption but the actual existence of distinct quanta of light that hung together like particles and did not spread like ripples on a pond. Einstein applied this idea to photoelectric emission, in which light ejects electrons from the surface of metals, to explain an experimental observation made a few years earlier by the German physicist Philip Lenard that could not be explained by a classical wave theory of light. Working at the University of Heidelberg, Lenard saw that as the light intensity increased, the energy of the ejected electrons did not go up as classically expected but instead only the number released per second increased. But the energy of the electrons did increase in direct proportion to the frequency of the light used, or as the color of the light was changed towards the blue end of the spectrum.

This was consistent with what Planck's findings about the discrete emission process for light had indicated, and what a quantum theory of discrete photons that contradicted classical electromagnetic wave theory required. The ratio of the energy of photons to their frequency was a new constant of nature, which became known as Planck's Constant.

Einstein's theory for the photoelectric effect and thus the existence of particle-like quanta of light was later confirmed in detailed experiments by Robert Millikan, the American physicist who won the Nobel Prize in 1923 for this work. In addition to his studies of the photoelectric effect, Millikan had also carried out a long series of painstaking experiments which showed that electrons always carry exactly the same fundamental charge, another form of quantization that was not predicted by classical electromagnetic theory.

Using these ideas on the discrete or quantized character of the steps in the emission of photons, Bohr had formulated his enormously successful "planetary" theory of the hydrogen atom that explained the puzzling regular patterns of light of different colors or frequencies that had been known for decades to be emitted by hydrogen. Bohr's postulate of a stable circular orbit, whose angular momentum was assumed to be given by the value of Planck's constant divided by 2π, also explained the size of the negatively charged electron's orbit around the positively charged proton at the center. Bohr's assumption of quantized angular momenta thus explained the amazing stability of the whole system, which on the basis of classical electromagnetic theory should have radiated away the electron's energy, leading to its rapid collapse into the proton. And this would have made it impossible to understand the existence of the hydrogen atom.

In 1927, Heisenberg showed that, in view of the existence of Planck's Constant as proven by the enormous success of Bohr's theory of the atom, Einstein's theory of the existence of quanta of light, and de Broglie's theory of the wave-nature of the electron, there was an inherent limit in an observer's ability to simultaneously determine both the speed and location of a particle. The energy of the photon used to detect the location of the particle in a microscope was determined by Planck's Constant in such a way that the more compact a photon or quantum of light was, the more energetic it had to be. This was shown by Millikan's experimental proof of Einstein's photoelectric theory: photons of higher frequency and shorter wave-length had increasingly more energy. Thus, the particle one was trying to locate was increasingly disturbed as one tried to determine its position with greater precision.

In principle, Heisenberg demonstrated that it would be impossible to predict the future course of events with perfect or infinitely great certainty as had been assumed in classical physics, as the French mathematician Pierrre Simon Marquis de Laplace had suggested in the early nineteenth century. A hypothetical observer could not determine the simultaneous momenta and positions of all the particles in the universe with infinite precision in order to predict its detailed future. But, to most scientists, the end of classical determinism also seemed to indicate that it would forever be impossible to devise detailed mechanical or geometric models for phenomena on the atomic or even smaller nuclear scale—such as that of the neutron.

Athough Heisenberg had originally proposed a theory of nuclear forces based on the exchange of electrons between neighboring neutrons and protons, he argued that the neutron was simply a neutral state of a massive class of nuclear particles or nucleons (the common name adopted for protons and neutrons). In a manner that remained unspecified, a neutron would somehow turn into a proton, changing its charge and mass, but retaining the strength of its interaction with nearby nucleons.

A purely abstract description of the neutron seemed to be necessary, because de Broglie's matter waves existed even for neutral particles such as the neutron, and not just for charged particles such as the electron. Accordingly, Heisenberg and Born's view that the de Broglie waves did not represent a physical entity—that they were purely formal constructs that determined the probability of finding a particle at a certain location—became widely accepted. Thus, the electron was regarded as an infinitely small point-particle whose fuzzy or indeterminate location was mysteriously determined by these "probability waves." Nothing further was needed in order to succesfully make all calculations, according to this view.

Having to regard the electron at one moment as a point particle and at other times as if it were a wave-like entity, together with the apparent impossibility of obtaining more than statistical descriptions of atomic phenomena, led Niels Bohr to formulate the idea of "complementarity." Bohr contended one simply had to accept the mysterious wave-particle duality as an inherent feature of phenomena on the atomic scale, which were governed by abstract quantum laws of probability without any detailed space-time description or deeper understanding, while on the large scale of ordinary phenomena, the laws of classical mechanics applied. For Bohr, Heisenberg's Uncertainty Principle indicated that the attempt of an observer to gain information on the microscopic scale inevitably involved disturbing the processes to such a degree that one could not form a clear picture. Bohr accepted the idea that there simply was nothing more to be understood than the mathematical formalism and how to apply it in practical calculations as developed by Heisenberg and Born.

This view came to be known as that of the Copenhagen School, since Heisenberg, who strongly believed that atomic phenomena are inherently not visualizable, often visited Bohr at his institute in the capital of Denmark to discuss things. Soon they attracted a large number of bright young physicists from all over the world. But Einstein refused to sacrifice the detailed understanding of phenomena on the microscopic scale in favor of abstract, probabilistic formalisms, which were only useful for the calculation of a few practical things like radiation emission, absorption and scattering processes. These abstract theories failed to provide a comprehensible picture of atomic phenomena.

As the philosopher of science Karl Popper put it, Einstein belonged to a small minority of theoretical physicists who kept insisting that one must try to interpret the quantum formalism, to understand it physically in the hope of gaining a better understanding of its physical difficulties and shortcomings, and thus eventually to use physics for explaining the cosmos.

The debates between Bohr and Einstein on whether the quantum theory as developed by the Copenhagen School was the final word continued for years, with neither side conceding defeat. The problem of how to reconcile Einstein's geometric space-time approach to physics, as in his General Theory of Relativity, with the abstract algebraic and probabilistic approach of Heisenberg, Bohr and Born to quantum phenomena remained unresolved in the lifetime of these scientists. So too did the nature of the ultimate, indivisible constituents of matter in the universe, crucial for an understanding of how the universe began.

CHAPTER 3

FROM X-RAYS TO MUONS

THE CONFIRMATION OF THE NEUTRON'S EXISTENCE and the need to postulate the neutrino undermined the hope for a classical, purely electromagnetic description of matter. The discovery of X-rays and other penetrating radiation in the environment, together with the evidence for so-called cosmic rays reaching the Earth from outer space, appeared to end this possibility altogether.

The discovery of X-rays by Wilhelm Conrad Röntgen in Germany at the end of 1895 shook the world as few previous scientific findings have done. Röntgen showed that the X-rays allowed an observer to see through the human body, an astonishing development that was immediately put to practical use by physicians all over the world. Within a few months, what appeared to be similar penetrating radiation from a uranium-based mineral was discovered by Antoine Becquerel in Paris. A number of other scientists immediately began to investigate the various sources of naturally occurring radiation in the environment besides uranium.

By early 1898, Marie Curie had found an element other than uranium that emitted penetrating radiation, namely thorium. Joined by her husband Pierre, who was the director of laboratory studies at the School for Industrial Physics and Chemistry in Paris, she discovered two previously unknown new elements by the end of that year. The first of these the two scientists named polonium, after Poland, the country of Marie Curie's birth, while the second, even more intense emitter—of which they were able to obtain only the tiniest trace—they named radium.

This last element promised to be so important as a source of powerful radioactivity, a term they proposed for this new phenomenon, that the Curies decided to devote themselves to the enormously laborious effort of extracting a significant amount of it. In 1902, after three and a half years of painstaking work with tons of Czech pitchblende ore, laboring under primitive conditions in a ramshackle wooden shed without a floor and a leaky roof in the yard of the school, they succeeded in chemically separating a tenth of a gram of radium. For this Herculean effort they were awarded the Nobel Prize the following year, sharing it with Becquerel. But because Pierre Curie had become ill from the prolonged exposure to radiation and they were both overworked, they could not come to Stockholm to accept the prize in person. Tragically, three years later, their fruitful collaboration came to a sudden end when Pierre Curie was killed in a traffic accident. He was 46.

The enormously concentrated source of powerful alpha particles that the Curies had produced allowed Rutherford to discover the planetary structure of the atom with a massive, positively charged nucleus at its center and an electron circulating around it. He noticed that on rare occasions the fast-moving, heavy alpha particles, with a mass equal to that of some seven thousand electrons, when fired into thin foils of heavy atoms would be deflected at a very large angle, rather than being simply slowed down. Only a close collision with a small but highly massive charge could produce such a sudden large deflection in a heavy projectile like the alpha particle. This was totally unlooked-for, on the basis of the then-favored model of the atom developed by the director of the Cavendish Laboratory at Cambridge University, J. J. Thomson, to which Rutherford had come as a young research fellow in 1895 from New Zealand.

In Thomson's model, the positive charge, together with the very low mass electrons, was assumed to be uniformly distributed throughout the atom, so that no sudden, large deflections of the massive alpha particle should occur. In one of his books, Gamow later compared this unexpected backward scattering of the alpha particles by atoms to the case of a customs officer shooting a bullet into a bale of cotton and finding it coming right back out at him, clearly indicating a concealed hard object like a gun. The lowest mass of all Rutherford models, hydrogen, had only a single positively charged proton at its center, and one negatively charged electron moving around it. It was for this simple system that Niels Bohr had devised his theory for the allowed orbits and the light it emitted when the electron dropped from outer to inner orbits in 1913.

In 1919, when the end of World War I allowed Rutherford to resume his researches, use of the same intense natural source of alpha particles discovered by the Curies led Rutherford to realize the dream of the medieval alchemists of transforming one chemical element into another. Bombarding the nucleus of nitrogen, Rutherford disintegrated it and obtained oxygen and a proton. Thirteen years later, his student James Chadwick used a radium source of alpha particles to prove the existence of the neutron. Thus the labor of the Curies was destined to bear enormous fruit in physics as well as in medicine, where the much more penetrating gamma rays emitted by radium were soon used in radiation therapy to destroy cancer cells.

The discoveries of Becquerel and the Curies made it clear that there were widespread and previously unsuspected sources of radiation in the environment. A number of scientists began to investigate this new phenomenon using a simple instrument called an electroscope to detect it. This instrument made use of the formation of ions or electrically charged particles that Thomson and Rutherford had found to be produced in the air as a result of charges torn off from neutral molecules by X-rays. It consisted essentially of two strips of gold-leaf suspended from a metal rod in a glass container, the gold leaves held together at one end so as to form a V-shape when negatively charged, due to the repulsion between the charges. When the molecules of air in the container have electrons removed by the radiation, as in the photoelectric effect at much lower energies, the resulting positive atoms or ions neutralize the negative charges placed on the gold leaves, causing them to slowly come together. From the speed with which this happens, it is possible to measure the intensity of the radiation.

In 1912, one of these investigators working in Austria, Victor Hess, took such an instrument in a balloon in the expectation that when it was high enough above the ground, the intensity of the radiation would decline. Instead, to his surprise, at an altitude of 15,000 feet the intensity of the radiation increased as much as nine times over that on the ground. Although others had noticed radiation increased with altitude, they attributed the rise to possible radioactive gases in the air. Hess was the first to argue that the radiation came from outer space.

In the mid-1920s, Millikan confirmed Hess' theory with extensive observations, and named this form of natural radiation "cosmic rays." Millikan believed that the cosmic rays were powerful forms of electromagnetic radiation consisting of gamma ray photons, or an electromagnetic form of radiation like X-rays. Gamma rays were the most penetrating of the three types of radiation given off by radium, the others being the heavy alpha particles that were stopped in a few thousands of a millimeter in solids, and the low mass beta particles, subsequently proven to be fast electrons with intermediate penetrating power of a few millimeters. Gamma rays had been found to be emitted from radium and other radioactive elements by the French physicist Paul Villard in 1900, but the ones Millikan believed to be of cosmic origin carried vastly more energy.

Millikan believed for a long time that these cosmic gamma rays originated in the outermost regions of the universe where matter was still being created, or that it was what he called "the birth-cry of matter." As has been pointed out by Isaac Asimov, Millikan was one of the few scientists who actively fought to reconcile religion and science. He was the son of a Congregational minister and deeply religious himself. The thought that matter was still being formed by a Creator had deep religious significance for him: "The Creator is still on the job," he said. And so he clung to the idea that the cosmic rays were primarily gamma rays emitted by the creation of matter in distant parts of the universe, even when more and more evidence suggested that they consisted for the most part of very energetic protons that produced the ionization of the air.

Millikan's view of the origin of cosmic rays apparently influenced the young Belgian priest Georges Lemaître. Studying astrophysics first at Cambridge University and then at M.I.T. in the mid-1920s, Lemaître developed his ideas of galaxies and stars born in the explosive, radioactive-like decay of highly compact fractions of an original "primeval atom" that gave rise to an expanding universe, accompanied by powerful gamma rays.

Millikan began his most famous work, on the electron's charge as the smallest in nature, at the University of Chicago in 1906. Initially, he had observed the motion of small electrically-charged droplets of water falling through the air with gravity's force, against the pull of a charged metal plate above. By 1911, he had switched to oil drops, which had the advantage of not evaporating while being tracked with a microscope. By exposing the apparatus to X-rays, now and then a charged particle would become attached to a slowly falling droplet. This caused the droplet to slow down abruptly or even to rise under the action of the electric field produced by the charged plate. The minimum change in velocity Millikan attributed to the addition of a single electronic charge, and by balancing the effect of the electrostatic attraction upward against that of the gravitational attraction downward before and after such an addition, he was able to calculate the charge of a single electron.

But what was so particularly important in Millikan's measurements was that he could show that the electric charge existed only as a whole number of units of a certain magnitude of the charge, and never as a fraction of this "natural minimum" value. This was a direct confirmation of the particulate nature of electricity first suggested by Benjamin Franklin, whose scientific work in the eighteenth century showed lightning to be an electric discharge. The discrete charge concept was further developed by Faraday in the next century through his experiments with electrolysis, in which a current is passed through a liquid containing salts, but a universal minimum value of electrical charge was not predicted by Maxwell's theory of electromagnetism based on Faraday's researches.

However, such a universal smallest possible value of the electron's charge was strongly suggested by J. J. Thomson's work. Thomson had become the director of the Cavendish laboratory at Cambridge University after Maxwell's death in 1879. Using a cathode ray tube very much like a modern television tube to produce a stream of electrons, and using magnetic and electric fields to deflect the stream, Thomson measured the ratio of the electron's charge to its mass with high precision in April of 1897. From these experiments, Thomson deduced that the electron had precisely the same but opposite charge of a hydrogen atom that was stripped of its negative charge, but that its mass was some two thousand times less. Moreover, he was able to show that the electron had exactly the same mass and charge in different chemical atoms, as had also been argued earlier that year by the German physicist Walter Kaufmann. The first sub-atomic particle common to all forms of matter had finally been proven to exist.

Although a few other scientists believed from their studies that there was no unique, minimum charge and that smaller charges appeared to exist, Millikan's conclusion was strongly supported by Bohr's theory of atomic structure, published only two years after Millikan's first oil-droplet experiment in 1911.

In 1923, Millikan was awarded the Nobel Prize for this enormously important work. The Nobel Committee also cited the crucial experiments that Millikan had carried out on the photoelectric effect, in which Einstein had postulated the quantized nature of electromagnetic radiation (and for which he had been awarded the Nobel Prize two years earlier). Millikan's researches had established experimentally both the classically unexpected "quantization" of an electric charge and the particle-like nature or "quantization" of light, the foundations of modern quantum theory. His work had therefore shown in the most direct and dramatic way that the supposedly indivisible "atoms" of hydrogen, regarded as the constituents of all chemical elements, were in fact themselves composed of two much smaller, more fundamental units of matter that shared a common absolute value of electric charge and appeared to differ only in their mass: the electron and the proton.

But Millikan's contributions to our understanding of the nature of matter and its origin in the early universe did not end there. When he moved to the California Institute of Technology from the University of Chicago to work on cosmic rays, he laid the foundation for the discoveries of his student Carl Anderson. In 1931, while studying cosmic radiation, Anderson found what came to be known as the positron, the first of many new particles of matter. The positive "twin" of the electron, the positron turned out to have the same mass as the electron, and a charge exactly equal and opposite to it as well. The precise value of this fundamental charge had been measured in years of painstaking work by Millikan.

Anderson had used a cloud chamber placed into a powerful magnetic field to visualize the tracks of particles produced by cosmic rays. The cloud chamber was a device originally developed by the Scottish meteorologist Charles Wilson to study cloud formation, to duplicate in the laboratory the effect that had intrigued him as a young boy on Mount Nevis, the highest peak in Scotland. By 1895, Wilson was working at Cambridge in the laboratory of J. J. Thomson, allowing moist air to expand in a container. The expansion lowered the temperature so that not all the moisture could be retained indefinitely in vapor form, and as a result the excess came out as water droplets to form a fine mist or cloud. The droplets normally formed around dust particles in the air, but he found that in dust-free air, they would form around ions or charged molecules such as were produced by the newly discovered X-rays.

While experimenting over many years, Wilson observed that fast-moving charged particles such as electrons and alpha particles from the newly available radium sources left tracks of clearly visible water-droplets when illuminated against a dark background. These could be used to identify the particles by the way they curved in the presence of a magnetic field, just as Thomson had originally measured the mass of the electron. Positively charged particles would curve in one direction, and negatively charged ones in the other, the lighter particles being bent more strongly than the heavier ones, and the more slowly moving bent more strongly than the faster ones. Also, the density of the tracks was greater for the more massive alpha particles than for the much lighter and faster moving electrons, and one could see collisions with molecules and other particles very clearly.

Rutherford had discovered the proton in 1911 by counting the flashes of light from individual alpha particles on a fluorescent screen. Subsequently, Wilson's cloud chamber became an increasingly useful way to study the visible phenomena of the microscopic nuclear world and to record it with a camera for careful measurement. For this pioneering work, the forerunner of the greatly improved cloud chambers utilized in nuclear particle physics for decades to come, Wilson received the Nobel Prize in 1927.

In 1931, at the suggestion of his mentor Millikan, Anderson used the cloud chamber, incorporating the strongest magnetic field ever used, to study cosmic rays. Anderson noticed that there were many curious, curved tracks in his photographs that indicated a positive charge if they were moving downward or a negative charge if they were moving upward. But in discussing this phenomenon with Millikan, Anderson noted that they could not be protons coming down, because their tracks were too thin, like those of electrons. "Everyone knows that cosmic ray particles go down," Anderson said. So he placed a lead plate into the middle of the cloud chamber to distinguish between upward- and downward-moving particles. The lead plate would cause a charged particle to lose energy in passing through it, so that by recording the curvature of the track before and after it passed through the plate in the presence of the strong magnetic field, Anderson was able to distinguish the direction from which the particles came. This simple idea allowed him to prove that some of the particles that produced curved tracks were those that would be expected from a positively charged electron coming down, and not a negatively charged electron going up.

The existence of such an "anti-electron" had been suggested just a few months earlier in a paper by Paul A. M. Dirac in England, who had earlier identified it with the massive proton. As Anderson recalled many years later, he had heard vaguely about Dirac's rather esoteric paper, but had been too busy with his experimental work to read it, and the discovery was therefore quite accidental. The editor of *Science Newsletter*, to whom Anderson had sent the first picture in December of 1931, suggested naming the particle the positron, and although Anderson did not like it, this name has been used ever since.

With the positron, there was now another particle of matter of the same mass and charge as the electron, but there was apparently no place for it in normal matter since it combined with an electron in a small fraction of a second, giving rise to pure radiation, usually in the form of two powerful gamma rays. Dirac, who shared the 1933 Nobel in physics with Erwin Schrödinger, had suggested the possibility of "anti-matter," in which the massive, positively charged proton would be replaced by its negatively charged twin, the so-called anti-proton, while the orbiting electrons would be replaced by positively charged "anti-electrons." But it would be decades before a few anti-protons were finally found in cosmic rays and produced by particle accelerator experiments. There was wide speculation that entire galaxies made of anti-matter existed, but the failure to find more than a few anti-protons among the cosmic rays argued against it.

In 1935, only three years after the discovery of the positron, the British physicist Patrick Blackett, who had studied under Rutherford, discovered that positrons could be produced together with an electron in a pair when high energy gamma rays disappeared after entering a lead plate. This was the first clear-cut observation of the conversion of energy into matter according to Einstein's famous relation $E = mc^2$, representing the opposite process in which an electron-positron pair annihilated into gamma rays. It was also support for the old idea first proposed by Lorentz at the beginning of the century that the electron and possibly all matter might be purely electromagnetic, since there was no "residue" of any non-electromagnetic form or "ponderable" matter left over when the electron and positron formed gamma rays.

However, the discovery of the neutron and the need for a neutrino, together with the strong and weak nuclear forces that could not be related to electromagnetic theory, made this idea obsolete. Instead, most theorists had begun to treat the electron mathematically as an infinitely small point, using newly developed ways to nullify the resulting infinite electromagnetic mass from a "ponderable" mass to obtain the observed value by various subtraction procedures in a theory called quantum electrodynamics.

The idea that all of matter might be nothing but a form of localized electromagnetic field energy was given an even more powerful blow the following year. In November 1936, Anderson and another former student of Millikan's, Seth Neddermeyer, published the first brief announcement of the discovery of an entirely new kind of charged particle, which appeared to have a mass between that of an electron and a proton. Soon, others confirmed this startling finding.

Because the mass of the new particle was between that of the electron and the proton, the authors suggested the name "mesotron," based on the Greek word for intermediate, but this was soon shortened to meson. Although most of the physics community was unprepared for such a particle, a young Japanese theoretician by the name of Hideki Yukawa had been thinking about the problem of the strong nuclear force since beginning his academic career in 1932, when the neutron was discovered and Heisenberg published his first paper on this subject. As Abraham Pais describes it in his history of particle physics *Inward Bound,* Yukawa at first thought of this force as being similar to that between two protons in a hydrogen molecule that has lost one of its two electrons, where the remaining electron is exchanged between the two massive protons in a kind of game of catch.

This was essentially how Heisenberg had approached the problem, but after a conversation with Yoshio Nishina, the founder of experimental nuclear studies in Japan, who suggested to him that perhaps the "ball " in this game might be of an entirely new kind with a spin similar to that of a photon or twice as large as that of the electron, Yukawa began to think about a new type of force. Unlike the electric force, it would have to have a very short distance over which it could influence another particle, based on the scattering experiments of Rutherford and others in the years since the proton was first discovered. But according to de Broglie's and Schrödinger's conclusions on the wave-aspect of particles that the size of a region in which a particle can be confined is inversely related to its mass, Yukawa deduced that such a particle producing the new short-range or nuclear force would have to have a larger mass than the electron.

A simple calculation showed that for a range of action comparable to the spacing between neutrons and protons in the nuclei of atoms, the mass of this "heavy particle" as he called it, would have to be about 200 times larger than that of the electron. He published this result in a Japanese journal in 1935, just a year before a new particle of about this mass was discovered. But it was not until two years later that anyone in a Western publication mentioned Yukawa's ideas, when Robert Oppenheimer, together with his associate Robert Serber, wrote a paper suggesting that the meson discovered by Anderson and Neddermeyer might be Yukawa's "heavy particle."

As it turned out, it took another ten years to learn that the cosmic ray particle discovered by the two students of Millikan was not actually Yukawa's meson. In early 1947, experiments by the British physicist Cecil Powell and his group in Bristol found that cosmic ray particle tracks produced in photographic emulsions suddenly came to an end and showed a kink that indicated another charged particle of mesonic mass was formed.

It was soon confirmed that there were indeed two kinds of mesons, one that was heavier with a mass of about 270 electron masses and given the name of pion, and a second one to which it gave rise, the muon, with a mass of about 200 times that of the electron. The pion seemed to fit the requirements of Yukawa's theory for the particle whose exchange between protons and neutrons provided the strong nuclear force. But the one into which the pion decayed, the meson discovered by Anderson and Neddermeyer, did not. In fact, it interacted very weakly with nucleons, the collective name adopted for the heavy neutron and proton, and seemed to behave just like an electron, interacting with matter by ordinary electromagnetic forces. The muon behaved as if it were just a short-lived, heavier version of the electron that it gave rise to when it decayed about a millionth of a second after it was produced in the disintegration of a pion.

The existence of two different kinds of mesons, one of which fitted theoretical expectations about nuclear forces while the other seemed to have no useful role to play, was a very puzzling new aspect of nuclear phenomena. As told by Abraham Pais, who was present when Powell reported this discovery in Copenhagen in September 1947, Powell began his talk with the following quotation from Maxwell: "Experimental science is continually revealing to us new features of natural processes and we are thus compelled to search for new forms of thought appropriate to these features."

It was the effort to solve the puzzle posed by the existence of these unstable new cosmic ray particles and their relationship to the stable electron and proton that kept me occupied for decades. This growing interest in the nature of the fundamental particles caused me to speak with Einstein in the spring of 1947, when an unexpected opportunity to meet him because of my work on electron emission suddenly arose.

CHAPTER 4

EINSTEIN

WITH A GREAT DEAL OF TREPIDATION I walked down Mercer Street in Princeton on a beautiful cool spring day. I was looking for number 112, the address that had been embossed on the top of the letter with the invitation to see Albert Einstein at one o'clock in the afternoon.

Here I was, in my early twenties, without any advanced education in physics, about to ask the most renowned scientist in the world since Newton what he thought about my ideas. The ejection of electrons from solids under bombardment by fast electrons was a subject I had just begun to study in my new job at the Naval Ordnance Laboratory, after having been discharged from the Navy less than a year earlier. I knew that this process was only loosely related to the photoelectric effect for which Einstein had won the Nobel Prize, and my ideas on the fundamental problems of quantum and particle theory that I had briefly referred to in my letter were still undeveloped. And so I was a little uneasy as I slowly mounted the steps to the porch of Einstein's house and rang the bell.

The door was opened by a middle-aged woman who I thought was probably his secretary. She asked me to wait in a room just to the left of the entrance, where there was a dining room with a large table in the center and chairs along the wall. Looking around the room, I was surprised to see a number of what appeared to be religious icons on the walls. After a few minutes, I heard soft steps along the hallway, and there was Einstein, exactly as I had seen him so often in recent photographs, with long white hair standing up like a halo around his head, dressed in an informal, baggy gray gym suit and wearing slippers.

Speaking softly with a German accent, he said, "Let's go to the back porch and talk there. It is such a nice day."

The covered porch of the simple white clapboard house faced a large lawn surrounded by bushes, old trees and spring flowers, and after we sat down looking out on the garden, he picked up a long white clay pipe and filled it slowly. After lighting it, he turned to me with a smile to ease my obvious anxiety and said, "So tell me about your ideas on electron emission."

Slowly regaining my composure, I started to talk, but he stopped me.

"Tell me, do you still speak German?" he said.

"Yes." I still had a rather heavy accent.

"Well, then," he replied. Let's talk in German. It's easier for me. Go on, tell me about your ideas."

Thus began a conversation that was to last an incredible five hours, a conversation that ranged from the details of the mechanism of secondary electron emission to the philosophical differences between the Copenhagen approach and Einstein's view of the incompleteness of quantum theory, from the nature of photons to the frustration of trying to find a unified theory of matter, and from the details of electron models to personal advice on how to avoid having all original ideas crushed out of me by going back to school for a doctoral degree.

I started to explain that I had been asked to examine the possibility of developing a new television technique for night vision by detecting the normally invisible infrared radiation emitted from various objects. It occurred to me that this might be achieved through the action of the infrared photons on the electrons emitted from crystalline substances by reducing the distance that they could travel as the crystal was being heated by the incoming radiation. These so-called secondary electrons would be produced by scanning the surface with energetic electrons as was being done in certain types of television camera tubes that had already been developed. And so, I had to learn everything I could about the phenomenon of secondary electrons being ejected from solids that had been discovered by the German physicist Philipp Lenard. It so happened that Einstein had developed his theory of photoelectric emission by discrete quanta of radiation on the basis of another set of experiments involving the ejection of electrons from metals by light, also carried out by Lenard.

Now and then Einstein would interrupt me with a question or a comment, but encouraged by his surprising interest, I went on to tell him how I found that the existing theories for this phenomenon discovered more than fifty years ago by Lenard did not agree very well with observations in the case of metals that had been studied most often. I had learned that the latest theory, based on an idealized theoretical model, had assumed that only the interaction of the incoming electrons with the most loosely bound ones needed to be considered. These were the same electrons involved in the theory of the photoelectric effect that Einstein had developed in 1905. But unlike the case for the low energy quanta of light that Einstein dealt with, the energy of the incoming electrons was much greater, enough to knock out the more firmly bound electrons in the deeper lying shells of the atoms in the metal.

Einstein nodded his leonine head. "That sounds reasonable to me," he said. He went on to point out that an idealized quantum mechanical model of metals appropriate to explain electrical conduction or the photoelectric effect would indeed not be appropriate for the relatively high energy electrons striking the metal. He added that too often, oversimplified mathematical models have limited applications in new physical situations.

Thus encouraged, I went on to outline my thoughts on how a better theory might be developed for the emission of secondary electrons.

When I had finished, he commented that I seemed to be on the right track, and that it was important to keep pursuing my ideas. I felt very pleased, and not a little surprised by the evident interest that he had shown in this rather detailed, technical subject related to practical applications of physics, when his own work for decades had been dedicated to a highly mathematical approach to a unified theory of electric and gravitational phenomena. I did not realize at the time that he had kept up an interest in the applications of physics to practical problems throughout his life, an interest which he had developed in his early years as a patent examiner. Only much later did I learn that he had been working with others on such practical problems as how to improve refrigeration and obtain better X-ray images using lead grids to reduce the effect of scattered radiation.

"Now let's talk about some really interesting things. Let's talk about the other subject you mentioned in your letter," he said, "the problems of quantum theory and the fundamental particles of matter."

I began by telling him that I felt that of all the particles that had ever been discovered, only one appeared to act like a truly fundamental particle, which seemed to be the ultimate, smallest charge with the smallest measured mass, and that was the electron. It had exactly the same but opposite charge of the much more massive proton, which also appeared to be stable, but with its charge concentrated in a much smaller physical volume. But apparently the proton was a more complex particle since it also possessed a short-range force that seemed to involve the newly-discovered meson. However, the mesons discovered in cosmic radiation had been observed to give rise to ordinary electrons or positrons, and so it seemed logical to find physical models that would somehow explain the mesons in terms of the stable particles. Furthermore, I told Einstein that I was struck by the fact that no one had ever seen an electron and a positron annihilating with any "ashes" or signs of any "ponderable matter" left over. Thus, they turned into nothing other than pure electromagnetic energy, two gamma rays, or two quanta of electromagnetic radiation.

Einstein nodded, indicating that he had thought about this often himself, but that the whole problem of the origin of mass had remained a puzzle, and that the present theoretical efforts in quantum theory that treated the electron as a point source of an electrostatic field led to infinite masses, requiring questionable mathematical subtraction procedures to arrive at the experimentally observed values.

I continued by saying that according to a series of recent theoretical papers dealing with the way fast electrons interact with atoms, there seemed to be no difference between the fields surrounding electrons and quanta of light in this respect.

Einstein agreed that the field of a moving charge, which flattens as a result of its relativistic contraction in the direction of motion, would excite or knock out electrons from an atom in the same way that a photon would. And then he asked me a question that took me quite by surprise, coming from the man who had suggested that discontinuous quanta of light or photons must exist: Did I have any idea as to what a photon might look like?

I did not have an answer, but I said that somehow it must bear a strong resemblance to the flattened field of a fast-moving electron. Einstein then asked whether I could explain the evidence that photons are emitted in a specific direction from an atom, unlike the case of a radio antenna, where the emission of electromagnetic waves takes place symmetrically in two opposite directions. Again, I had to admit that I could not explain this deviation from classical electrodynamic theory, but that this sort of "needle-like" aspect of photon emission in a single direction was something that I hoped could eventually be understood in geometric and dynamic terms.

Einstein then brought up another conceptual problem involving the nature of photons. This was the question of how one can understand that a single photon, which acts like a point particle when it knocks out an electron from an atom, can also act like a wave when it creates an interference pattern as it passes through a screen with two widely separated slits in it, so that the photon appears to be in two different places at the same time.

This problem was at the heart of Bohr and Heisenberg's argument that quantum theory required one to give up the effort of arriving at detailed space-time models for phenomena on the atomic and nuclear scale. It was similar to the problem posed by de Broglie's theory of matter waves for the electron, namely how a wave-like interference pattern could be produced when electrons are fired against a screen with two slits one at a time, seemingly requiring that the point-like electron had to be in two places at once.

It was in large part this conceptual difficulty that had persuaded me already as an undergraduate to regard the electron as a pure, stable source of an electromagnetic field extending over large distances without any "ponderable" hard core of some unknown form of matter, an idea that Einstein evidently had not been able to accept completely. Nevertheless, in our discussion, he did not discourage me from pursuing this idea, and as will be discussed later in this book, ten years later it was my paper on this subject that resulted in de Broglie inviting me to work on an electromagnetic description of matter at the Institute for Theoretical Physics in Paris.

After about an hour or so, Einstein's secretary came to tell him that there was someone who had an appointment, and I thought this would be the end of our conversation. However, to my surprise, he turned to me and said that I should wait here on the porch. After a few minutes, he was back.

Returning to the question of the stability of the electron and its relation to the other particles of matter, I suggested that one could perhaps make progress in understanding the newly discovered particles—since there was no evidence that the electron ever breaks up into smaller charges—by for the time being not trying to solve the problem of what holds the electron together. Why not just accept it as a stable source of an electromagnetic field of finite size, while trying to understand the neutron and the mesons without introducing any unknown type of ponderable matter or new forces, and assume that the mass of the electron is simply due to the energy in its field?

Einstein replied that he had himself at various times considered the electron as a purely electromagnetic entity and at other times as one that contained ponderable matter, but that he had been unable to settle the problem in his own mind. Also, he could see no reason why in principle one could not conceive of charges smaller than that of the electron. He agreed that one probably has to accept the idea that the electron has a finite size, but that there were theoretical difficulties in the treatment of a finite-size electron in relativity theory that needed to be worked on.

Einstein then asked me how I would treat the question of the stability of atoms that Bohr had answered by postulating that in the smallest orbit, or normal state, the electron does not radiate, contrary to classical physics, a problem that I had been thinking about as alluded to in my letter. For the smallest circular orbit in the hydrogen atom, Bohr had assumed that the angular momentum, or the product of its mass, its radius and its velocity, is given by Planck's Constant *h* divided by 2π, the ratio of the circumference of a circle to its radius. Bohr never gave any physical explanation for the failure of the electron to radiate and spiral into the proton at the center when given this particular angular momentum. Many theorists before and after him, including J. J. Thomson, had tried to solve the problem of the stability of atoms in terms of Maxwell's classical electromagnetic theory, and at that time, I was hoping to find an explanation in recent calculations of Peter Debye and Julius Stratton about oscillating but non-radiating electrically charged spheres.

Einstein was very emphatic that it was hopeless to use the classical theory of Maxwell to resolve this problem, although he agreed that one must try to develop detailed descriptions of phenomena on the atomic scale that were *anschaulich* or visualizable. He said that the view of Bohr and Heisenberg that one cannot ever hope to do this was wrong, and that to be satisfied with merely formal mathematical and statistical predictions of the outcome of experiments would never lead to explanations of the fundamental problems of physical theory. He added that although he had as yet no alternative for existing quantum theory, he was sure that the purely probabilistic view was inadequate, although he doubted that he would live long enough to confirm this view.

After about another hour, his secretary came to tell him that there was something else he needed to take care of, and once again he told me to wait a few minutes. When he came back, he suggested that we take a walk in the garden, where our conversation took a more philosophical and personal turn.

"You see the large tree over there," he said. "Now turn your head away. Is it still there?" At first, I was puzzled about what he was driving at, but it soon became clear. He was explaining to me one of the principal aspects of the Copenhagen interpretation of quantum theory that he found particularly unacceptable, according to which an observation or measurement is necessary to bring an object like an electron into definite existence. In this view of atomic phenomena, since one only has a probability distribution for the location or velocity of an electron, given by the de Broglie wave function as interpreted by Heisenberg and Born, it is only when an observer makes a measurement that an object or an event becomes real.

It was this mysterious, almost magical aspect of the quantum theory that Einstein told me he abhorred, an approach that he deeply felt to be a great mistake, an opinion that I told him I completely shared. The Copenhagen view of quantum theory was in consonance with the positivistic philosophy that had become increasingly accepted in the latter part of the nineteenth century, favored in particular by the Austrian physicist and philosopher of science Ernst Mach and the philosophers that shared his ideas in Vienna. Mach strongly opposed the use of unseen and insensible objects to explain phenomena, and was an opponent of the atomic theory, since there were no direct observations that proved their existence at the time. All knowledge for Mach was a matter of sensations supported by measurements, as he wrote in 1872, and he maintained that the laws of nature were just convenient, man-made generalizations to describe empirical observations, but that it was only these many quantitative observations or measurements themselves that had reality, a view strikingly similar to that adopted after Mach's death in 1916 by Heisenberg, Bohr and Born.

As Einstein explained his views to me, he said that in his early years he had been strongly influenced by Mach's insistence that theories should be based entirely on empirically derivable or directly observable quantities, but later he concluded that this was a mistake. Particles like electrons and alpha particles exist before they produce a click in a detector or a flash on a fluorescent screen, just like the tree existed before I saw it.

He went on to say that he no longer believed in some of the fundamental ideas of his Special Theory of Relativity that were the result of Mach's early influence on his thinking. It was only subsequently that I understood what Einstein was referring to, namely his assumption at the time he worked on the Special Theory that there was no such thing as an ether, since it could not be directly observed or detected by any physical measurements. The inability to detect the so-called luminiferous ether postulated by Maxwell and others as the medium in which electromagnetic waves are propagated had been discovered in the experiments of Michelson and Morley in the 1880s. These experiments had failed to detect a motion of the Earth through the hypothetical ether. Einstein added that he still had a high regard for Mach, but that he had to reject some of his ideas in formulating the General Theory of Relativity.

Again, I was not sure exactly what he had in mind, and only many years later did I realize that this remark was related to the universal space-time continuum that he had to introduce into his theory and whose curvature gives rise to gravity. Such a medium was a concept that resembled in certain aspects the ether of Descartes, Newton and Maxwell that Einstein rejected in his special theory of relativity, under the influence of Mach's views.

As Einstein explained in a lecture at the University of Leyden in 1920, "To deny the ether is ultimately to assume that empty space has no physical qualities whatever. The fundamental facts of mechanics do not harmonize with this view." In particular, he cited Newton's conclusion that rotation of an object in empty space, which influences its mechanical properties, has to be taken as something real, so that Newton might well have called his absolute space "ether." Thus, by 1920, Einstein had completely rejected Mach's hypothesis that there is no ether and that the resistance of objects to a change in their state of motion or their inertial mass was due to the effect of distant masses in the universe. However, as Einstein put it, the ether of general relativity is not a uniform medium as conceived in the wave theory of light. Instead, it is at every place conditioned by the presence of matter at a particular location and in neighboring places.

I was surprised to find that as our conversation continued into the late afternoon, Einstein spoke about how he had become increasingly discouraged about his work, repudiating much of what he called his early "rash" and "false" ideas. "Something completely new has to be found," he said, "something that is somehow based on the ideas of General Relativity." And so it was a shock to hear him refer to some parts of his 1917 paper, in which he applied his General Theory of Relativity to arrive at the ground-breaking model for a closed universe—the beginning of modern cosmology—as no longer of any significance. He had earlier regretted introducing a repulsive force of unknown physical origin into his equations in order to obtain a universe that would remain stable and not collapse under the influence of gravity, characterizing it in the early 1930s as one of his greatest blunders. It was this decision, consistent with the absence of observational evidence for anything but a static universe at the time, that had prevented him from predicting an expanding universe. Many years after our encounter, I realized that this was after all not a blunder, and that the mysterious force he had postulated to oppose gravity was indeed necessary even in an expanding universe, and that it arose naturally in a rotating universe.

As we returned to the house, Einstein asked me to tell him a little about how I had become interested in physics and about my plans for the future. I told him that in my childhood, my father, who was a dermatologist, had talked over the dinner table about X-rays and their use and misuse to my mother who was a gynecologist and pediatrician. I mentioned how they had always tried to the best of their ability to answer my questions about X-rays, the radio waves produced by the diathermy or heat-producing machine, and the ultra-violet lamp that were in my father's office. I also mentioned how they bought me popular books about physics and simple experiments that had fascinated me when I was child.

I told of how my father had once taken me to the laboratory of a patient of his in Berlin who was the inventor of the yellow sodium-vapor lamps on our highways, which impressed me enormously. I went on to tell him of the curious coincidence that as a medical student at the University of Heidelberg my father had received his education in physics from the lectures by Philipp Lenard, who had been a student and later an assistant of Heinrich Hertz. Lenard had discovered the phenomenon of secondary electron emission. Hertz was the discoverer of both the radio waves that had been predicted by Maxwell and the photoelectric effect.

When I was still living in Berlin, I told Einstein, my father had urged me to become a physician, but all through my childhood I was more interested in learning about the physical nature of matter, radiation and the universe. Thus, when it came to registration at Cornell, I decided against the pre-medical curriculum, but compromising with the wishes of my mother that I pursue a practical profession that would give me a better chance to earn a living, I settled for electrical engineering with the idea that eventually I could take up physics in graduate school.

Thereupon Einstein asked me whether I intended to continue working at the Naval Ordnance Laboratory or whether I planned go back to school for an advanced degree in physics, and I answered that I was thinking about graduate school.

I was absolutely astonished to hear his reply. He urged me to remain in my present job doing applied physics and not to go back to school for a doctoral degree, where he suggested that I would have any originality crushed out of me. He asked me to think about this very carefully, not to go into the academic world of teaching, to study on my own what was of interest to me, and to work on my ideas about the fundamental problems of physical theory in my spare time so that I would not have to worry when I came up without any results for long periods. Thus, I would be able to make my mistakes in private.

I could hardly believe what I heard next. "Don't do what I have done," Einstein said. "Always keep a cobbler's job where you can get up in the morning and face yourself that you are doing something useful. Nobody can be a genius and solve the problems of the universe every day. Don't make that kind of mistake. When I accepted the position at the Kaiser Wilhelm Institute and the University of Berlin, I had no duties really. Nothing to do except wake up and solve the problems of the universe every morning. Nobody can do that. Don't make the mistake I made."

It was late in the afternoon when he took me up to his study overlooking the garden and pointed to the papers on the table. They were densely covered with equations, evidently his present work on the unified field theory that he had been working on for years. "You see all this ?" he said to me in a sad, dejected voice. "I will never know whether any of this has any meaning whatsoever."

It was six o'clock in the evening when I left to take the train back to Washington, deeply shaken by my experience. As I wrote down some notes on what had happened, I could not get over the realization that here was a man, widely honored and world-famous for making incredible advances in understanding our world and for the numerous humanitarian causes he helped so freely, clearly feeling a deep sense of failure and sadness towards the end of his life.

I knew some of the reasons that could explain his state of mind, including the enormous burden created by the atomic bombs that had killed hundreds of thousands in just two explosions and which now seemed to threaten the very survival of human life on Earth. He had always been a dedicated, outspoken opponent of all wars. The atomic bomb was a weapon for which he had taken a personal responsibility, since it was he who agreed to sign a letter prepared at the urging of Leo Szilard and Edward Teller, advising President Roosevelt about the possibility of an atomic bomb, which led Roosevelt to begin the program that eventually led to its construction.

After the atomic bombs had been dropped on Hiroshima and Nagasaki and the war had ended, Einstein had publicly warned that the Russians would soon be able to build their own bomb, and that this would lead to an arms race whose outcome would most likely be the destruction of human life on a much vaster scale than what had already been seen in two World Wars. I also knew that he had become involved in efforts with other scientists to warn the public of the full danger of nuclear war and the need to prevent a potentially suicidal nuclear arms race.

I could imagine how the revelations about Hitler's death camps, in which so many of his family and friends had died, must have affected him as a survivor, something that I could well understand, having lost my grandfather and all his sisters in the Holocaust while I had been able to reach safety in the United States with my brother and my parents.

And there was Einstein's recent isolation from the scientific community that largely regarded his almost lone opposition to the highly successful quantum theory as unrealistic and anachronistic. Most younger physicists tended to regard as hopeless his efforts to find a theory that would unify the forces of nature by relating electromagnetism to gravity. This was especially the case since the discovery of the neutron and its decay with the apparent need for a neutrino, together with the mesons found in cosmic rays, which seemed to require forces other than those that could be explained by electromagnetism or gravitation.

Later, I learned about other, personal reasons for Einstein's overwhelming sadness: the death of his wife only a few years after their escape from Hitler; the long mental illness of his younger son in Switzerland; his long illness; and the stresses of his life and the ceaseless efforts to help others escape the Nazi threat.

Apparently, Einstein also urged others not to pursue a career in higher education. As described in a 1965 article by Martin Klein, a professor of physics and historian of science who later became the senior editor of Einstein's papers, even in his early years in Berlin Einstein had told his friend Max Born not to worry about placing a student in an academic position. "Let him be a cobbler or a locksmith; if he really has a love for science in his blood and if he is really worth anything, he will make his own way."

As Klein tells it, Einstein's unhappy experience as a student at the Polytechnic Institute in Zurich, followed by the extraordinarily fruitful years as a patent examiner, when he had no academic connections whatsoever but during which he arrived at his most important insights, made him write the following: "It is little short of a miracle that modern methods of instruction have not already completely strangled the holy curiosity of inquiry, because what this delicate little plant needs most, apart from initial stimulation, is freedom; without that it is surely destroyed. . . ."

According to Klein, towards the end of his life, Einstein not only did not regret his lack of an academic post in his early years. He actually considered it a real advantage.

> For an academic career puts a young man into a kind of embarrassing position by requiring him to produce scientific publications in impressive quantity—a seduction into superficiality which only

> strong characters are able to withstand. Most practical occupations, however, are of such a nature that a man of normal ability is able to accomplish what is expected of him. His day-to-day existence does not depend on any special illuminations. If he has deeper scientific interests he may plunge into his favorite problems in addition to doing his required work. He need not be oppressed by the fear that his efforts may lead to no results.

Klein summarized Einstein's view of physics as a drama of ideas and not a battery of techniques. There should be an emphasis on the evolution of ideas, on the history of our attempts to understand the physical world so as to give us some perspective and to realize that, in Einstein's words, "the present position of science can have no lasting significance."

These were precisely the views that Einstein had expressed to me, and they were to have a major influence on my future thinking.

CHAPTER 5

THE BODY ELECTRIC

BUOYED BY THE ENCOURAGEMENT OF EINSTEIN, in the days and weeks after my return to Washington I pursued the development of a theory of secondary electron emission and my plans for experimental work on this phenomenon. The conversation with Einstein strongly supported my feeling that the principal problems of fundamental physical theory were of a conceptual and not a mathematical nature.

Einstein had struggled in vain for more than two decades to relate electromagnetic and gravitational phenomena primarily by mathematical means that became increasingly removed from the physically visualizable approaches with which he had made his most fruitful contributions in his early years. But he had encouraged me to pursue the attempt to find models for the particles of light and the particles of matter as extended, geometric and dynamic structures—contrary to what was believed possible by the Copenhagen School of quantum mechanics represented by Bohr, Heisenberg and Born. That attempt eventually led me to a theory of the origin of matter and an expanding, ultimately stable universe, of the sort Einstein had postulated on the basis of his General Theory of Relativity in 1917.

This turn of events was set in motion by the coincidence that in Washington, where the Naval Ordnance Laboratory was located, George Gamow was teaching a course on relativity and cosmology. Gamow had developed the theory of alpha particle emission from radium and other heavy elements. Although Einstein had urged me not go to graduate school, I felt I had to increase my knowledge of nuclear particle theory, relativity and cosmology by taking graduate courses at George Washington University, where I had accepted a part-time position as a teaching assistant in the physics department in the summer of 1946.

I was fortunate in being able to attend Gamow's lectures. One of the most original and gifted teachers of physics at that time, he was just then working on a new theory of the formation of the chemical elements at the beginning of the expansion of the universe, a theory that was later to become known as the Big Bang model.

Early in his life, Gamow developed a great interest in Einstein's theories of relativity and a closed universe. After completing his undergraduate studies in Odessa, Gamow went to the University of Leningrad for graduate work, where he began to study under Aleksander Friedmann, the first scientist to show that Einstein's equations for the closed universe would allow expanding and collapsing models of the cosmos. In 1922, Friedmann had published a paper in which he concluded that it was not necessary to assume that there was only a static solution to Einstein's equations, achieved by introducing an arbitrary repulsive force that exactly balanced the pull of gravity. After Einstein initially wrote an article attacking Friedmann's paper that the size of the universe could vary with time, a concept that he disliked, Einstein wrote a second note acknowledging that Friedmann's solution was in fact correct.

Gamow had begun his graduate work under Friedmann's guidance, but Friedmann died in 1927, so Gamow was forced to abandon cosmology for his thesis. After post-graduate fellowships in Göttingen with Max Born, in Copenhagen with Niels Bohr, and in Cambridge with Ernest Rutherford, where he worked on a theory for the production of the sun's heat by thermonuclear reactions, the brilliant twenty-seven-year-old returned to Leningrad in 1931 to accept a professorship in physics.

By then Gamow had already developed a successful theory of alpha particle emission on the basis of quantum theory published when he was only twenty-four, had written the first textbook on nuclear theory, and had developed the theory of the nucleus as a liquid drop that was used in 1939 by Bohr and Wheeler to develop a detailed theory of nuclear fission. However, in the early thirties, the political situation in the Soviet Union became increasingly oppressive under Stalin. Even in the field of physics Einstein's theories were not allowed to be taught, as they were regarded as inconsistent with dialectical materialism. Gamow decided to leave the Soviet Union. After a nearly fatal attempt to escape across the Black Sea to Turkey in a kayak, he finally managed to defect with his wife after a conference in Brussels, eventually ending up at George Washington University in 1934. There, in collaboration with Edward Teller, who himself had left Germany in 1935, Gamow continued his work on the theory of nuclear forces and nuclear processes that produced the heat of the stars by the fusion of hydrogen into helium. As it turned out, the forces at work in the hot interior of stars were the same that were harnessed by Teller and Ulam in the early 1950s to power the hydrogen bomb.

In World War II, Gamow worked on the development of the atomic bomb as well as on various naval ordnance problems (in the course of which he frequently visited Einstein in Princeton). By 1946, he had concluded it was impossible to explain the abundance of helium in the universe if this element were formed from hydrogen only in the interior of stars. Having studied with Friedmann, who had first shown the possibility of a universe expanding from an initially very dense state, Gamow and one of his part-time students—Ralph Alpher, who was working at the Applied Physics Laboratory in nearby Maryland—began to examine the possibility that helium and some of the other elements came from the fusion of hydrogen in an extremely hot initial state of the universe.

In developing this idea quantitatively together with Robert Herman, a colleague of Alpher's who was expert in the mathematics of general relativity, Gamow and his collaborators calculated the abundance of helium and other light elements formed in the first few minutes after the formation of the neutrons in the early universe. In this brief period, the temperature of the rapidly expanding and cooling fireball was still very high, but not so high that the newly-created nuclei of atoms would be immediately destroyed. The required temperature was found to be a few million degrees, the same temperature found in the center of stars like our sun. From a knowledge of the amounts of the various elements existing now, the scientists were able to calculate what the temperature of the radiation emitted by the initial fireball would be today, after billions of years of cooling as a result of the expansion of the universe.

In the papers Alpher and Herman published within two years after Gamow gave Alpher the topic for his thesis, they concluded that the temperature of the radiation produced in the early universe must have declined by now to just five degrees Centigrade above absolute zero. This is amazingly close to the best value of 2.73 degrees measured today. Moreover, Gamow and his collaborators were able to predict the number of photons that had been produced relative to the number of nucleons existing at the time of the Big Bang, and thus the number per cubic inch that should exist today everywhere in the space around us.

However, it was not until 1964 that Robert Wilson and Arno Penzias, working at the Bell Telephone Laboratory in New Jersey, accidentally discovered these relics of the hot Big Bang as background noise in an antenna designed to pick up shortwave radio signals bounced off a satellite. As described by Rocky Kolb in his recent book *Blind Watchers of the Sky,* even then the origin of the background radio noise did not become clear until six months after the discovery. Wilson and Penzias happened to learn about a search for the relic photons of the Big Bang just then being started by James Peebles, a physicist at nearby Princeton University, who had given a talk about his plans at the Applied Physics Laboratory. Peebles himself was unaware that Gamow and his associates had predicted the existence of this radiation almost two decades earlier, and that this work had been done in the very laboratory where he gave the talk about his planned study.

The discovery of the cosmic background radiation, as it is now called, turned out to be the first observational evidence other than the redshift of the galaxies that supported the Big Bang model for the origin of the universe as developed by Gamow. It was a crucial discovery for which Penzias and Wilson received the Nobel Prize. As Kolb noted some three decades later in his account of these events, many scientists immediately realized that Gamow and Lemaître's ideas should have been taken more seriously. Even to this day the prescient work of Alpher and Herman, initiated by Gamow, has not been properly recognized.

Perhaps one reason why Gamow's ideas were not taken seriously enough was his flamboyant personality. He rode motorcycles and drove flashy convertibles to scientific meetings. He also wrote popular books with humorous illustrations about basic physics and astronomy that made him famous as an effective interpeter of the complex and mysterious world of relativity and quantum mechanics. These books, with titles such as *Mr. Tompkins in Wonderland,* were filled with amusing illustrations that he had largely drawn himself. They were similar to the popular book by the German physicist Paul Karlson that had attracted me to physics as a young boy.

In 1947, when I was attending Gamow's lectures, he had just published *One, Two, Three, Infinity: Facts and Speculations of Science,* with its account of his version of the Big Bang. It differed from Lemaître's version in that Gamow's account of the origin of the expanding universe began only after neutrons had been formed to create an extremely hot and dense fireball with a radius of hundreds of millions of miles. Lemaître had envisioned an earlier, unknown form of even denser matter in the form of a single "primeval atom," from which ordinary particles like Gamow's neutrons arose only in the final stage of a long process of repeated division by two. Gamow also differed from Lemaître in his belief the universe went through periodic cycles of expansion and collapse. Lemaître favored the idea that the universe expands only once.

At the time I attended Gamow's lectures on relativity and cosmology, his ideas about the formation of the chemical elements in the early universe were being successfully worked out by Alpher and Herman. Thus it was not surprising that he was bursting with enthusiasm as he lectured on cosmology and relativity. He strode back and forth in front of the blackboard chain-smoking cigarettes, in his absentminded absorption occasionally putting the chalk in his mouth instead of the cigarette.

Although astronomy and cosmology had always fascinated me, at that time in my life I was most interested in the particle problem, particularly the nature of the neutron. Since it was an electrically neutral particle that had a finite mass and had been observed directly in the laboratory, unlike the neutrino which was thought not to have any mass and was therefore more like a photon or a form of radiation, the neutron seemed to be the single greatest obstacle to explaining all forms of matter as purely electromagnetic. If the neutron could somehow be explained in terms of the electron and the proton, it might be possible to achieve the kind of simplification and unification of all forces in nature that Einstein had hoped for and that, before him, Faraday had dreamed of. I had first learned of this quest from my course in the history of science that I had taken as an undergraduate at Cornell.

In studying relativity theory in greater depth, I became aware of the work of the Dutch physicist Hendrik A. Lorentz, who taught theoretical physics around the turn of the century at Leiden University. Lorentz, who was regarded by Einstein as the greatest living physicist of his age, had shown that the contraction of objects and the slowing down of clocks in relativistic moving systems could be explained by electromagnetic theory. Later, Einstein arrived at these same results by a different method. Lorentz dealt with the failure to detect an ether by suggesting that all particles, including the neutral ones, were purely electromagnetic in nature or composed of equal and opposite numbers of charges. Lorentz's view appealed to me very strongly, since it provided a physical explanation for the contraction of matter and the slowing of clocks. It was also consistent with my growing conviction that the masses of the electron, the proton and the neutron as well as those of the newly discovered mesons, together with the forces acting between them, should eventually be explainable in purely electromagnetic terms.

Lorentz and his student Pieter Zeeman also explored the connection between magnetism and light, and their conclusions further supported the electromagnetic interpretation of matter. Lorentz was the first to suggest that the atoms that emit light, regarded as electromagnetic waves similar to the radio waves discovered by Heinrich Hertz in 1887, were composed of charged particles that could be set vibrating. In 1896, the same year that Becquerel discovered radioactivity and Lenard and Thomson were establishing the mass and charge of free electrons,

Lorentz suggested that Zeeman place a light source in a strong magnetic field. If atoms contained charged particles whose oscillations gave rise to light, that should then affect the wavelength or color of the light emitted. This was indeed observed, and by 1902, with the existence of electrons in atoms strongly supported by the discovery of beta particles emitted from atoms in radioactive decay by the Curies and Rutherford, Lorentz and Zeeman were awarded the Nobel Prize in physics. The connection between magnetism and light was something that Faraday had hoped to find, but only the advanced technology available to Zeeman made it possible to confirm Faraday's intuition.

With the the latest experiments supporting the idea that neutral atoms are composed of charged particles, Lorentz published a series of papers summarized in a lecture given at Columbia University in 1906. Lorentz explained that the failure of the Michelson-Morley experiment to show any effect of motion of the Earth through the ether could be explained if indeed all forces are related to the electromagnetic ones, and all mass is purely electromagnetic, a conclusion that he refined in a book published in 1909. Thus, Lorentz was able to explain the physical origin of the contraction of objects in the direction of motion which the Irish physicist George Fitzgerald had advanced in the early 1890s as a way to explain the inability of the Michelson-Morley experiments to detect the ether. On the basis of electromagnetic theory, Lorentz established that no matter how fast an observer might be moving relative to the ether, light waves emitted in all directions will appear to have a spherical shape and thus seem to have the same velocity in all directions, independent of the motion of the source.

The situation was exactly the same for an observer in a uniformly moving ship, for whom the ordinary Newtonian laws of motion hold. The observer cannot detect the fact that he or she is moving by experimentation, such as by throwing a ball upward and having it come down to the same spot, as Galileo had first shown three centuries earlier. In fact, the recognition that objects could have two independent components of their velocity at the same time as that due to the movement of the ship and the vertical movement of the ball helped win acceptance for the Copernican idea that the Earth could rotate about its axis and move around the Sun without leaving debris thrown up behind. Lorentz's calculations showed that the same was true for all electromagnetic experiments involving light or radio waves, so that all the laws of physics hold for an observer on the Earth moving through the ether, even though relative to the ether the observer's lab might be traveling at nearly the speed of light.

Lorentz concluded that since the normally spherical electric field of a charge at rest would flatten to a thin disk and the size of its source shrink when it moved, its mass would also have to increase. In classical or Newtonian mechanics, the mass of an object is assumed to be independent of its speed. But when an electron approaches the speed of light, the volume of the source of the field as well as the thickness of its field approaches zero, and thus its field energy density and therefore its mass becomes infinitely large. It was another indication that the velocity of light is an upper limit to the speed with which any object can travel, whether charged or neutral. This result clearly finds its simplest explanation if all matter is in fact composed of nothing but charges whose mass is purely electromagnetic in origin. And since the many experiments in the early years of the twentieth century regarding the increasing velocity of electrons confirmed Lorentz's predictions, it seemed to him probable that matter could indeed be regarded as entirely electric.

But to the young Einstein pursuing his ideas at the patent office in Bern, this approach to resolving the failure of all experiments to detect motion through the ether seemed unsatisfactory. It was not at all certain that all forces are forms of electromagnetic interaction, or that all mass is due to the energy in the electromagnetic fields of charges. It seemed possible that there is "ponderable" mass in nature unrelated to electrical charges, and that there may be other than electromagnetic forces. In fact, at the turn of the century, chemical forces appeared to differ radically from electrostatic forces in their dependence on the separation between atoms. Also, the energies involved in radioactive decay were enormously greater than could be expected from the relatively low energies involved in chemical interactions, so there might be new kinds of forces involved. And efforts to find mechanical models for the propagation of electromagnetic waves in an ether had become extremely complex and had failed to explain all the properties of light. Electromagnetic or light waves were found to involve electrical fields that were at right angles to their direction of travel, which cannot be explained by a simple fluid model for the ether in which only compression waves can exist, as is the case for sound waves in air. Instead, Maxwell's equations seemed to require that the ether behaves more like an elastic body that is rigid, which was hard to reconcile with the apparent absence of any resistance to the motion of the Earth or the other planets through the ether.

As Einstein was considering the problems presented by the various mechanical models for an ether as the medium in which light waves are propagated, he learned of Lenard's 1902 experiments, in which light ejected electrons from metal surfaces. This so-called photoelectric effect seemed explicable to Einstein only if light came in bundles or individual photons that behaved like discrete particles and not at all like waves in an ether such as the radio waves that had been discovered by Hertz. Just as Democritus saw atoms as particles moving in a vacuum, such photons, regarded as packets of light that did not spread out as they traveled long distances, did not seem to need a medium, as for example do waves of sound. Thus, the evidence for the photoelectric ejection of electrons from metals whose energy did not depend on the intensity of the light waves falling on the metal surface, as would be expected classically, argued against light as spreading waves. Lenard had found that the energy of the ejected electrons depended only on the frequency of the light pulses, not on the intensity of the light. Only the number of electrons ejected depended on the intensity of the light, which could be understood if there were a larger number of particle-like quanta in an intense light beam. And all this could be explained only if light consisted of entities that had the properties of point-like particles, as difficult as this was to reconcile with the success of the theory of light as waves in describing numerous optical experiments and the spread of radio waves emitted by antennas.

Just a few years earlier, Planck was inspired by experimental results about the spectrum or the color of light emitted from a hot object as its temperature changed, showing that the emission of radiation was taking place in discrete or quantized steps. Now Einstein was forced by Lenard's results on the photoelectric effect to assume that the light emitted in such discrete steps would move through space not like waves but like particles of matter, the energy remaining confined in a fixed volume rather than spreading like a wave. And if indeed this conclusion was correct, then there was no need to assume the existence of an ether as a medium for the propagation of light.

When Einstein was thinking about the conceptual problems of light as particles or as waves in some sort of stationary fluid medium that would be a universal reference, he was deeply influenced by the ideas of the philosopher of science Ernst Mach. A convinced positivist, Mach held that scientists should not invoke any invisible, undetectable entities to explain physical phenomena such as the ether. Mach had even suggested that mass and thus the inertia of objects, or their tendency to resist a change in their state of motion, is due to the existence of distant masses in the universe. If there were no such distant stellar masses, Mach argued that an object would have no mass or inertia at all—clearly a concept of mass that had nothing to do with electromagnetic fields and was far more general.

Mach rejected Newton's conclusion that there must be an ether or absolute reference frame to explain why water in a bucket rises along the outer wall when the bucket rotates. For Mach, this could not happen if the bucket of water were the only object in the universe, since there would be no meaning to rotation and thus no centrifugal force in the absence of other masses relative to which rotation could be defined.

Following Mach's view of the origin of inertial mass, Einstein thought he should try to do without an ether or any absolute reference frame, so that there would only be relative motions and no absolute ones at all. He decided to derive the Lorentz-Fitzgerald contraction and the slowing down of clocks in moving system by a totally different approach. Instead of figuring these changes in the length of measuring rods and the rate of clocks from the theory of electromagnetic interactions, as Lorentz had done, Einstein followed the example of Galileo and boldly postulated that all the laws of physics, including those of electromagnetism, must be exactly the same for any observers moving with uniform straight-line motion relative to each other. Secondly, he postulated that the speed of light measured by any such uniformly moving observer relative to another is independent of the state of motion, and that there is therefore no preferred or absolute reference frame or ether.

Based on these assumptions, together with certain specific ways to define how distances and time intervals were to be measured by each observer using rigid meter rods and light signals, and with a careful analysis of how simultaneity of events can be determined by observers moving relative to each other, Einstein was able to derive the laws found by Fitzgerald and Lorentz without any reference to the nature of matter or the nature of the forces acting between particles, and without the need for an ether or absolute reference frame.

It was Einstein's very general derivation of the contraction of moving objects and the slowing down of clocks in moving systems that Gamow taught, and not the theories of Lorentz, who is not even mentioned in his popular books. But in hindsight, this is really not so surprising. Gamow did all his successful work on the basis of nuclear theory, where one experiment after another indicated that neutrons and protons interacted with forces some one hundred times stronger than the electromagnetic force, on the basis of which Lorentz had arrived at his results. Moreover, the strong nuclear force needed to explain all the recent experiments, as well as Gamow's own theory of the nucleus as a liquid drop, was of very short range. The nuclear force decreased with distance much more rapidly than the electrostatic force between charges—the Coulomb Force, that declined just like gravity with the square of the distance.

Finally, there was also the "weak" nuclear force that governed the emission of neutrinos in the process of beta decay that Enrico Fermi had developed in the early 1930s. That weak nuclear force accounted very well for radioactive beta-decay, just as did Gamow's quantum theory of alpha particle decay. Following a suggestion by Wolfgang Pauli in 1931 in an effort to solve the problem of apparently lost energy in beta decay, Fermi assumed neutrinos were particles without charge or mass. He postulated they were a form of radiation more weakly absorbed by matter than light, X-rays or gamma rays, explaining why they had not been detected.

Neutrinos were thought to be created when a neutron disintegrated into a proton and an electron, carrying off a part of the energy and a spin half that of all forms of ordinary radiation in order to explain the observed spin of the neutron. They seemed to be associated with a third force, unrelated to electromagnetism. As a result, for Gamow and most other theoretical physicists there appeared no reason to believe that all natural forces were forms of the electromagnetic interaction, as required by Lorentz's derivation of the equations for the shrinking of objects and the slowing down of clocks in moving reference frames. Einstein's more general approach that abandoned the idea of an ether did not need this assumption, and was therefore preferred.

But neither Fermi's theory for beta decay nor the existing purely phenomenological theories of the strong nuclear forces that Gamow and other theorists were developing accounted for the existence of two completely different types of mesons recognized in 1947. It seemed that every time a new phenomenon was discovered, scientists would have to assume that there was a new kind of particle and a new kind of force involved that had no connection with electromagnetic forces—and thus no hope for the kind of unified, rational understanding of physical phenomena that Faraday envisioned and Einstein kept hoping and working for. Moreover, so far every newly-discovered unstable particle that possessed a finite rest-mass eventually ended up emitting electrons or positrons.

I decided to keep searching for an electromagnetic model for the neutron and the nuclear forces along the lines envisioned by Rutherford, who had first proposed the neutron's existence as a new, compact type of state of the proton and the electron. Encouraged by Einstein to follow my own ideas, I diverged from Gamow, Heisenberg, Bohr and the vast majority of other physicists, who were convinced that trying to develop a detailed space-time model of the neutron was ruled out by the nature of the Uncertainty Principle and the firmly-established experimental knowledge of the spins of the nuclear particles.

Around this time, I was quite unexpectedly encouraged by a theoretical paper on the behavior of electrons passing through matter. In my research on secondary electron emission I had come across both it and a book on the history of ideas, brought to my attention by David Baumgardt at the Library of Congress, the same person who had suggested that I write to Einstein.

While studying how atoms in a solid determined how far electrons hitting the solid's surface would penetrate, and how the electrons scattered, I found a paper by C. G. Darwin on the interaction between very rapidly moving electrons and protons. A quite startling mathematical result was detailed by Darwin—the grandson of the biologist Charles Darwin. My Darwin was a mathematician at the University of Manchester who had helped Rutherford with the theory of the scattering of alpha particles. These experiments led Rutherford to conclude that the positive charges with most of the mass of the atom were concentrated in a small nucleus. Rutherford observed that a few of the massive alpha particles were unexpectedly scattered backwards from the very thin layers of gold and other substances they were hitting. Rutherford discovered the proton and determined its approximate size when he bombarded hydrogen with the positively charged alpha particles. That in turn led to the planetary model of the atom for which Bohr developed the theory of quantized orbits in 1913, after he had visited Rutherford for a few months in Birmingham the previous year.

The paper that was so surprising was one in which Darwin investigated the affect of the theory of relativity on the scattering of a negatively charged electron that was strongly attracted to a positively charged proton. To my amazement, I found that Darwin had discovered that under certain conditions, an electron approaching a proton would not be scattered but would instead be captured into a spiral orbit, bringing it close to the proton in a finite period. This would occur only if the angular momentum of the electron, or the product of its velocity, mass and radius was initially below a certain small critical value. It occurred to me that this process might be involved in the formation of a neutron, the particle predicted by Rutherford in the early twenties and only found by Chadwick in 1932.

It would be the pursuit of these unexpected relativistic spiral orbits found by Darwin that ultimately led me to electromagnetic models for the nuclear particles more massive than the electron, and decades later to a model for Lemaître's "primeval atom" whose division explained the origin of Gamow's hot fireball of neutrons and the highly ordered structure in the universe.

The ideas of Karl Joel, an Austrian historian whom Baumgardt had known as a young man, helped provide some perspective on my quest for a simpler, more unified description of the nature of matter. Joel's studies in the history of ideas convinced him of the existence of a cyclical pattern in which periods of unification and synthesis alternated with periods of analysis and the discovery of a multiplicity of new and unrelated phenomena. During the period of discovery, there was a revolt against unification and systematization, characterized by an anti-rational and mystical trend that influenced not only science but also art, literature and politics. Joel found that these views of the world changed radically in about three generations, or a span of some one hundred years from one extreme to another, completing a cycle in some two hundred years.

As Joel pointed out, while the middle of the nineteenth century had been a time of great unification in physical theory based on rather simple geometric, mechanical or fluid models as developed by Helmholtz, Faraday, Maxwell, Thomson and Lorentz, the early twentieth century saw the discovery of a multiplicity of experimental phenomena whose theoretical description did not seem to be visualizable or related to anything that had been understood previously. This new phase in physical theory began with Planck's discovery at the turn of the century that radiation is emitted from a hot object only in certain discrete amounts governed by a new fundamental constant of nature subsequently known as Planck's Constant—the beginning of quantum theory. Five years later, Einstein concluded that radiant energy was not only sent out in discrete amounts but that it also remained strangely localized, unlike waves which spread in a spherical manner, a view that had explained a wide variety of optical phenomena with enormous success.

Equally upsetting was that as a result of Einstein's theory of relativity, the widely accepted Newtonian idea of an ether as an absolute reference frame relative to which motions could be measured had to be abandoned. The ether had also provided a medium for the propagation of light. As described above, at least two new types of non-electromagnetic forces appeared to be necessary to explain the great strength of nuclear forces and the emission of electrons accompanied by neutrinos in radioactive beta-decay of nuclei.

All this happened only a few years after Bohr's radical atomic model discarded the idea that charges moving in circular orbits lose their energy by radiation, according to the laws of electrodynamics. Bohr's theory also required discontinuous jumps of electrons from one orbit to another, against classical laws demanding continuity in all natural phenomena. This development was followed by the yet more radical renunciation by Heisenberg of all attempts to find detailed geometric or mechanical models for phenomena on the microscopic level based on his Uncertainty Principle. The ability to predict the future course of events with unlimited accuracy, the basis of classical physics since the time of Newton, had to be abandoned.

Compounding the difficulties of a rational understanding of matter was the discovery that particles had wave properties and that these matter waves were not of a material nature. In some mysterious way, these waves merely determined the *probability* of finding a point-like particle in a certain place. Finally, the discovery of a growing number of totally new particles that appeared to have no relation to electromagnetic phenomena, beginning with the first meson in the early 1930s, seemed to end all hopes for a unified understanding of the universe.

This historical view of the cyclical changes of scientific ideas gave me hope that over the course of the years, a new era of unification of our theories on the nature of matter and the evolution of the universe would come about, and that it was most likely to involve electromagnetic concepts, as Lorentz had believed to the end of his life. But this development would take a long time, during which I would have to be able to earn a living in a "cobbler's job" while working on the theoretical and conceptual problems of physical theory on the side, as Einstein had urged me to do. I realized that I would need an advanced degree. I had loved living in scenic Ithaca with its deep gorges and waterfalls while pursuing my undergraduate studies at Cornell, and so I decided to return there while keeping my position at the Naval Ordnance Laboratory, coming back to Washington during the summers and using my work on secondary electron emission for my thesis subject.

CHAPTER 6

"BE STUBBORN"

IN EARLY 1949, I returned to Cornell for graduate studies in the newly formed program of Engineering Physics. The university had attracted some of the world's most outstanding scientists to its faculty, many of whom had been involved in the Manhattan Project that developed the atomic bomb. Among these was Richard Feynman, the brilliant, brash young theoretical physicist who had been instrumental in developing new methods of computing in Hans Bethe's theoretical division at Los Alamos. Bethe had been instrumental in bringing Feynman to Cornell shortly after the end of the war in 1945. Feynman shared an office with Philip Morrison, who had also worked on the atomic bomb at Los Alamos, and who agreed to serve as my principal adviser in theoretical physics on my thesis committee. Feynman sometimes joined my discussions with Morrison about the nature of the neutron and the mesons newly discovered in cosmic rays. This brief acquaitance led me, a decade later, to work out the mathematical details of an electromagnetic model for the mesons that turned out to be the basis for the Lemaître atom.

I had originally hoped to do my thesis work on the theory of secondary electron emission under Bethe. Bethe was most widely known for his work on the nuclear reactions in stars that produced the heavier elements from hydrogen, accounting for the source of their energy along lines similar to those studied by Gamow. Bethe had also written some definitive papers on energy loss of fast particles passing through atoms, elaborating on the work of C. G. Darwin and the pioneering studies of Niels Bohr. Secondary emission works like a cue ball striking the racked balls on a pool table: fast "primary electrons" strike the surface of a solid and eject other secondary electrons from the atoms they hit. Gradually, the primary or incident electrons lose energy. It thus seemed to me Bethe would be interested in serving as my thesis adviser, particularly since Einstein had encouraged me to pursue my ideas in this field. But when I talked to Bethe, he said that he had done some work on secondary emission and concluded that a theory of this complex phenomenon could not be readily achieved.

Although I was disappointed by Bethe's reaction, I remembered Einstein's warning not to allow myself to be easily discouraged, and I decided to find someone else to work with. Morrison and Lloyd P. Smith, chairman of the Physics Department and head of my thesis committee, were willing to let me go ahead, and allowed me to pursue experimental and theoretical work on secondary electron emission at the Naval Ordnance Laboratory. By the fall of 1950, I had found a very close relation between the number of secondary electrons ejected from metals and their position in the periodic table, showing clearly that the strongly bound electrons played a crucial role in the phenomenon, the idea Einstein had encouraged me to pursue. This permitted me to write my first scientific paper, published in the *Physical Review*. I sent a copy to Einstein, together with an essay on the concept of time that I had written in my course in philosophy of science taught by the philosopher Max Black, who had agreed to be on my doctoral committee for my minor subject in this area.

To my great delight, Einstein replied within a few weeks, saying that my results for the secondary electron emission process seemed to argue strongly for my hypothesis, and that my investigation of the concept of time appeared reasonable in many ways. Einstein did not quite agree with the importance I placed on Immanuel Kant's contribution to the origin of the concepts of space, time, and causality, which could not be derived by a logical process from empirical observations. Einstein argued that it was not true that there were certain *specific* concepts without which thought was impossible, as Kant had maintained. Rather, it was the case that without concepts in general, thinking was impossible. But exactly what the nature of these concepts were could not be known *a priori*.

My initial results on the mechanism of secondary electron emission were accepted for my master's thesis, and the detailed theory of the phenomenon that I was able to work out in the following two years formed the subject of my doctoral thesis. My work on secondary emission in the case of insulators such as potassium chloride, which I had begun to study at the Naval Ordnance Laboratory, later led to a job at the Westinghouse Research Laboratories. A new family of detectors in high energy accelerators, used to study particles similar to the mesons, eventually resulted from the phenomena involving this simple salt. Years later, thin foils carrying a layer of potassium chloride in powdered form were used to store electrical charges, making possible a new type of television tube to transmit ultraviolet images and spectra of stars from the first orbiting observatories in space. The same type of camera tubes—which I also worked on, decades later—allowed us to witness the first steps of a human on the moon as this historic event was actually taking place.

At Cornell, I worked under the guidance of Philip Morrison, a far-sighted and open-minded individual. Ten years or so younger than Gamow, Morrison shared with him an enormous enthusiasm and ability to make complex ideas vividly clear, both to his students and to the general public. Morrison had a wide interest in science, technology and the history of ideas reaching far beyond the field of theoretical physics, in which he had received his doctorate under Robert Oppenheimer, who directed the development of the atomic bomb. Morrison was willing to listen patiently to the unconventional ideas of some of his students, never discouraging them but always coming up with probing questions that touched the heart of the problem.

Morrison had been involved firsthand with the difficult and dangerous task of designing and assembling the plutonium cores of the first two implosion-type nuclear bombs at Los Alamos, and had also been among the first American scientists to visit Hiroshima just after the Japanese surrender. He was deeply committed to warning the public and politicians of the need to prevent another world war. As a result, he became one of the founders of the Federation of American Scientists, an organization dedicated to peace. Morrison wrote numerous articles to describe the horrors he had seen in Japan, not just those produced by the atomic bombs but also by the fire-bombing of dozens of cities. Over a hundred thousand people had been killed in thousand-plane B-29 raids. In this task he was aided by Einstein, who also used every opportunity to call for an end to nuclear bomb testing and the elimination of weapons of mass destruction. This aspect of Morrison's and Einstein's concerns greatly influenced my own life.

Morrison was also interested in astronomy and cosmology, areas in which some of his students were pursuing their graduate studies. One of these was Vera Rubin, who was examining the motions of some one hundred nearby galaxies. She concluded that they seemed to be rotating as a group, so that the universe as a whole might be rotating, just as the individual galaxies were rotating about their centers. This was regarded as a very controversial result. At the time, it was widely believed that the galaxies were uniformly distributed in an expanding space, so that the only motion should be one of recession as seen by us, with the farthest galaxies moving at the greatest speed, giving rise to a uniformly increasing redshift as distance increased. Morrison supported Rubin in this controversy, but when she presented her findings at a meeting of the Astronomical Society in Philadelphia in 1950, she was strongly attacked. She had such difficulty in getting her thesis work accepted by the astronomical community that she abandoned this particular area of research. She moved to Washington where her husband had accepted a position at the Applied Physics Laboratory run by Johns Hopkins University, the same laboratory where Alpher and Herman had worked on the Big Bang model, and she did her Ph.D. thesis on the clustering of galaxies under Gamow. A few years later, one of the most respected French astronomers, Gerard de Vaucouleurs (who had accepted a position in the United States in 1955), concluded that there was indeed strong evidence that the distribution of galaxies was not uniform or random. De Vaucouleurs suggested that the local galaxies Rubin had examined were part of a supercluster that did indeed appear to be both rotating and expanding. When I later arrived at a model for the universe in which there had to be even larger structures composed of many superclusters that also rotated, I realized that this model was supported by de Vaucouleurs' views. The universe appears to be a highly ordered hierarchical system composed of rotating systems of increasing size, as first envisioned by Immanuel Kant two hundred years ago, rather than a random collection of galaxies.

Rubin eventually became a widely respected astronomer, working at the Carnegie Institution in Washington, D.C. Using electronic image intensifiers tried out by her colleague Kent Ford as a way to improve upon the limited sensitivity of photographic film, Rubin pioneered the study of the rotation of galaxies. A decade after leaving Cornell, I had an opportunity to try out a new type of electronic image intensifier with Ford at the Lowell Observatory in Flagstaff, Arizona, based on the work on secondary electron emission from insulating crystals that I had begun at the Naval Ordnance Laboratory. By the early 1970s, the work of Rubin and Ford on the rotation of galaxies had provided the most convincing observational evidence that these gigantic systems of stars were surrounded by an enormous halo of invisible "dark matter" of an unknown nature, for which the electromagnetic theory of mass I had discussed with Einstein in Princeton and later with Morrison and Feynman at Cornell offered a possible explanation.

As controversial as the idea of rotating superclusters or even a rotating universe were when Rubin and I sat in Morrison's class on modern physics at Cornell, I learned of an even more controversial theory about the nature of the universe through a lecture by the visiting British astronomer Thomas Gold. Gold, who a few years later joined the university's Astronomy Department, together with Hermann Bondi and Fred Hoyle at Cambridge University in England did not like the idea of a Big Bang as a singular occurrence, as advanced by Lemaître and Gamow. Instead, in a series of papers published a few years after World War II, Gold, Bondi and Hoyle proposed the radical idea that the universe was in a steady state of constant expansion that had been going on forever and would continue expanding indefinitely. They based this concept on what they called the "Perfect Cosmological Principle," according to which the universe is not only the same everywhere throughout space in our epoch, but also that it is unchanging throughout time, despite the evident expansion as demonstrated by the growing redshift.

However, there was one major problem with the idea of a steadily expanding but unchanging universe. The theory required that as the galaxies move apart, new matter appear out of nowhere, uniformly throughout space, so that the amount of matter per unit volume, or its density, remains the same. Not only was there no known theoretical explanation for a process where matter had to suddenly appear spontaneously everywhere to make up for the matter lost by the continuous expansion, even at a very slow rate, but there could be no conservation of mass or energy in a universe with a constant ongoing creation process, which was particularly hard to accept if the universe should turn out to be closed.

Gold made a strong case for a steady-state universe that had no unique moment of creation. He argued that continuous creation was no more miraculous than a single creation event as postulated by the Big Bang model. But only a minority of physicists and astronomers were attracted to this idea. Subsequent discoveries disproved the steady-state theory by showing that the universe was in fact different in the past. However, since at the time it was proposed this theory made a definite prediction that could be tested by observations of the early universe from study of very distant objects, it stimulated the construction of more powerful optical and radio telescopes to test this idea. The theory had the appeal that it predicted a universe that would exist forever, whereas the Big Bang appeared to require a universe that would some day die, either by flying apart into dead cinders as the stars burned out or collapsing into a singularity in which all life would come to a fiery end.

An aspect—one that did not violate the laws of conservation of matter and energy—of the idea of new matter continuously appearing out of "nowhere" turned out to be important for me. The steady-state theory's idea that the universe go on forever was based on Lemaître's and Gamow's ideas, in combination with Rubin's and De Vaucouleurs' conclusion that the system of galaxies to which the Milky Way belongs seemed to be rotating.

The controversy at Cornell about Rubin's results was followed by one I caused in the physics department the next year. Experiments that I began in 1951 in pursuit of an electromagnetic model for the neutron showed that neutrons could apparently be formed from protons and electrons at very low energies, far below the energy predicted by the existing theory.

The idea for this experiment occurred to me in reading about the search for the neutron by Rutherford and Chadwick during the decade after Rutherford had first predicted its existence in the early 1920s. I found that Chadwick had at one time looked for these elusive neutral massive constituents of the nuclei of all atoms in a hydrogen-filled electrical discharge tube. This tube was similar to an old gas-discharge type of X-ray tube system I had seen in a basement laboratory of Lyman Parratt, one of the senior members of the physics department at Cornell, who had also worked on the atomic bomb project. Such a gas-discharge tube operates like a modern fluorescent lamp, being nothing much more than a tube filled with a gas at a low pressure to which voltage is applied at the ends where metal wires fuse into the glass, conducting electric charges that allow a current to pass through the gas. But instead of applying one or two hundred volts or so as do present-day fluorescent lamps, in the early years of X-ray studies such tubes were operated at many thousands of volts. Electrons emitted from the negative end, called the cathode, would be accelerated and strike the electrode at the positive end or anode, thereby generating X-rays. It was exactly so that X-rays were accidentally discovered by Wilhelm Conrad Röntgen, who used this type of gas-discharge tube in his laboratory at the University of Würzburg in Bavaria in 1895.

Based on the theoretical work of C. G. Darwin, the mathematician who worked with Rutherford when he discovered the massive proton in the nucleus of atoms, I speculated that on rare occasions an electron coming sufficiently close to a proton might be captured to form a neutron. Evidently, this idea had occurred to Rutherford and Chadwick in their search for the neutron, but they did not have a method for detecting them efficiently. Enrico Fermi's discovery of neutron-induced radioactivity made such research possible.

In 1934, Frederic Joliot and his wife Irène Curie, the daughter of Marie Curie, found that some elements could be made artificially radioactive by bombarding them with alpha particles so they would emit positrons. Fermi found that many elements could be made to emit beta rays or energetic electrons when exposed to neutrons. The beta rays were captured by the massive nuclei in the center of the atoms. Only a few months after having published his theory of beta decay, Fermi bombarded a series of elements with neutrons including the heaviest element, uranium, in the expectation that he could form new and even heavier elements. He thought he had found an element heavier than uranium, but in fact missed discovering that uranium bombarded with neutrons would, with the release of a great amount of energy, actually split up into two lighter elements that were radioactive. This process, now generally referred to as fission, led to the atomic bomb. It was discovered some four years later by the German scientists Liese Meitner and Otto Hahn in Berlin.

Among the most important discoveries made by Fermi was that when the neutrons were slowed down to very low velocities by a series of collisions with hydrogen and carbon, as found in paraffin wax, their chance of being captured in foils of these materials would greatly increase. This would result in short-lived radioactivity in such materials as silver and indium, which could then be detected by taking the foils of these metals to a sensitive Geiger counter where the resultant beta and gamma rays could be measured. It was this technique of induced radioactivity that I decided to use for the detection of neutrons produced in Parratt's gas X-ray tube after persuading Parratt and my advisors to allow me to try this experiment. I argued that since I could use my work on the theory of secondary electron emission for my thesis, failure of the experiment to produce neutrons would not endanger my receiving a doctoral degree.

According to the existing theory of the neutron and its decay into a proton, an electron and a neutrino as worked out by Fermi, there was no chance that such an experiment could possibly succeed. The neutron was believed to have a mass so large that it would take an electron accelerated to about 780, 000 volts to produce it. But the power supply of Parratt's X-ray tube would only provide about 35, 000 volts, some twenty-two times less. Nevertheless, the neutrino had not yet been directly observed to exist, and it was possible neutrons did not all have exactly the same mass under all conditions. C. G. Darwin's calculations indicated that neutrons might be formed by capturing an electron even at low energies.

The likelihood that I would be able to produce neutrons was very small, and in retrospect it was amazing that I was allowed to carry out the experiment at all. It was only the open-mindedness of Morrison and Parratt that made it possible. And so, when after the very first few experiments with paraffin blocks piled around silver and indium foils close to the old brass X-ray tube showed signs of radioactivity thirty to fifty percent above the normal background of the detector, many of my colleagues could not believe this was due to neutrons being formed from low energy electrons and protons.

As I excitedly explained in a letter that I wrote to Einstein at the end of August 1951, the results could not be explained by the action of cosmic rays forming neutrons, because none were detected when no voltage was applied to the tube. It could also not be explained by contamination of the metal electrodes in the tube. I had replaced materials that could give rise to neutrons with newly machined parts. The possibility that a small normal admixture of deuterium, a form of heavy hydrogen, was the source of neutrons was eliminated both theoretically and subsequently by deliberately adding known amounts of this gas and measuring the neutron production rate. To everyone's consternation, no one in the physics department was able to suggest a known nuclear reaction that might explain the observed activity.

I mentioned in my letter to Einstein that in order to improve the ability to detect neutrons, two of the faculty, Giuseppe Cocconi and Kenneth Greisen, had offered to take the indium foils with their longer-lived activity of 54 minutes to a nearby salt mine, where they were carrying out cosmic ray experiments and the background count rate was much less than I was able to produce with six inches of lead to shield my counters in Rockefeller Hall. This was in fact done a few weeks later, although the first effort to speed the process of getting the foils to the counters two thousand feet below the surface by dropping them between two pieces of plywood ended in minor disaster when the foil crumpled upon hitting the ground. Chastened, we used the elevator to get the foils down to the counters in the mine over the next few weeks, and the evidence for neutrons being formed in the discharge tube continued to show up.

A few days after I sent my letter to Einstein, a reply arrived that did in fact contain a possible explanation of my anomalous result with rather disheartening implications for my attempt to do without the neutrino. After pointing out that an electron would have to acquire an energy of 780,000 volts to form a neutron, Einstein suggested that perhaps more than the energy produced by the applied potential might become available if more than one electron were to give up its energy to a proton at the same time, something that is conceivable according to quantum theory. He ended his letter by saying that since the results of the experiments were clearly important, further pursuit of the method would be necessary. He also raised the question whether it might not be advantageous to use an electron beam of known energy, and let it fall on a solid target such as paraffin that contained hydrogen so that the energy of the electrons could be brought under better control.

I answered Einstein and explained why I thought that even at relatively low energies neutrons might be formed if they had slightly different masses, an idea that I felt had not been completely ruled out. I did in fact follow Einstein's suggestions, further experimenting with the gas-discharge tube at Cornell for a few more months. I presented my results at a meeting of the Journal Club of the Physics Department later that fall. The improved sensitivity by doing the measurements of the indium foils in the salt mine continued to indicate the production of neutrons, but no one could find any explanation other than that suggested by Einstein.

By the end of the fall semester, it became clear that I had to give up my efforts for the time being in order to finish my work on secondary emission so as to get my doctoral degree.

Two years later, the experiment was independently repeated by Edward Trounson, a physicist and friend of mine at the Naval Ordnance Laboratory, with similar results. But when, some nine years later at the Westinghouse Research Laboratories, I finally had an opportunity to carry out the experiment with a separate electron beam interacting with both solid and gaseous targets in the form suggested by Einstein, no neutrons were produced.

To this day, just exactly how neutrons can be formed at much lower energies than expected in the complex environment of a gas-discharge tube remains a mystery. Neutrinos were finally detected in 1956 by Frederick Reines and Clyde Cowan. Since then, they have been observed in high energy accelerators, coming from the Sun and more distant stars, but there remains an unresolved puzzle about their production deep in the interior of stars. In the course of thirty-six years of experiments since 1960, despite increasingly sophisticated experiments, only half as many neutrinos have been observed as would be expected on the basis of theories for the Sun's energy production. Since the conditions in the interior of the Sun are somewhat analogous to those in a high voltage hydrogen discharge such as I used at Cornell, there may be some surprises awaiting us relating to the so-called fusion processes that produce the energy in stars involving the formation of helium from hydrogen, in the course of which both neutrons and neutrinos are produced.

Although the physical mechanism by which neutrons were formed in my experiments remained unexplained and my findings were a source of great controversy, I was not discouraged in my efforts to pursue electromagnetic models for the neutron and the many new particles then being discovered. A series of large nuclear particle accelerators went into operation at various universities during the three years of my graduate work at Cornell. And my studies in the history and philosophy of science convinced me that a period of great confusion, when many new and unexpected phenomena are discovered, is followed by a period of synthesis, when connections between apparently disparate phenomena are found. All the new unstable particles eventually decayed to stable charged particles such as protons and electrons accompanied by photons and neutrinos that could only exist while traveling at the speed of light, so the possibility of an electromagnetic nature of all matter and a composite character of the neutron remained the simplest hypothesis in my mind.

In fact, Fermi, who had developed a theory of the weak nuclear force in which neutrinos were involved, had published a paper together with the physicist C. N. Yang in 1949 in the *Physical Review*, the year I arrived at Cornell. Fermi and Yang suggested that the pi-mesons might be composed of protons and anti-protons, an idea that ultimately led me to a closely related model for the neutral pi-meson and still later to a model for Lemaître's primeval atom.

In early 1951, I submitted an article for a contest sponsored by the Institute for the Unity of Science. I outlined my ideas on the need to search for a unified approach to the conceptual problems of modern physical theory that would allow a more comprehensible, visualizable view of the nature of matter than that provided by the existing highly abstract mathematical approach of quantum theory as developed by Heisenberg and Born. Percy W. Bridgman, an eminent physicist and philosopher of science, was one of the judges. Although I did not win the contest, Bridgman invited me to visit him at Harvard. He told me that he thought my ideas were interesting, but that they needed to be better developed and tested by future experiments before they would find wide acceptance. Just as Einstein had urged me to do, he advised me to be patient, to keep pursuing my ideas on the conceptual problems of physical theory privately while doing other work in physics, and to be willing to wait until they had matured, perhaps for many decades, allowing me to make my mistakes in private.

In the spring of the following year, I gave a talk on my work on electron scattering and secondary emission at a meeting of the Physical Society. At the end of the lecture I was approached by a researcher working at the Westinghouse Research Laboratories in Pittsburgh, who asked me whether I would be interested in continuing these studies in connection with their potential application to electronic X-ray image intensifiers. His name was Milton Wachtel, and he told me of the ongoing efforts to reduce the necessary dose of X-rays required in medical fluoroscopic examinations. He and his colleagues were working with large vacuum tubes. Electrons released by the X-rays fell on a thin layer of luminescent material mounted on a thin sheet of glass inside the tube, producing photons that in turn ejected electrons from a thin metal film on the other side of the glass by means of the photoelectric effect. These photoelectrons then impinged on a luminescent phosphor screen after being accelerated by an applied voltage of many thousands of volts. The result was an image some two hundred times brighter than the luminescent screen used by Röntgen and further developed for fluoroscopic X-ray examinations by Thomas Edison at the turn of the century.

This sounded like an interesting application of my work in electron physics to the obviously important problem of reducing the risk of cancer from diagnostic X-rays. I visited Pittsburgh, where a brand new research laboratory was about to be completed. When I was told that I would be able to pursue my interests in nuclear particle physics as well as finish writing my thesis while equipment to continue my experimental studies was being built, I accepted the offer and moved to Pittsburgh in the summer of 1952.

The opportunity to carry out useful and interesting research of a practical nature while pursuing my work in theoretical aspects of physical theory was exactly the kind of non-academic position Einstein had urged me to find. In March of 1954, I sent Einstein a copy of an abstract of another paper in the field of secondary electron scattering by solids that had just been published, together with a letter congratulating him on his 75th birthday. I received what was to be his last note to me before his death in 1955. Written across the back of a printed thank-you note were the following two words: "Be stubborn."

CHAPTER 7

TWO PLACES AT ONCE

INSTEAD OF FINISHING MY THESIS on the theory of secondary electron emission while I was at Cornell, I had focused on whether neutrons could be formed in a high voltage gas discharge. I used most of my time at the Westinghouse Research Laboratory to complete my thesis, and obtained my doctoral degree in 1953. The ongoing efforts at Westinghouse to reduce the dose in X-ray fluoroscopy and devise a television tube operating in the ultraviolet were exciting opportunities to put my work in electron physics to practical use. The X-ray work had the potential to save many lives; the television tube was destined for telescopes orbiting the Earth as part of the Harvard-Smithsonian Astrophysical Observatory.

Although I had to put aside the effort to devise and test a model for the neutron, I kept working on the relation between matter and light in my spare time. A flood of new particles was being discovered with the aid of the large accelerators then coming into use, producing a growing crisis in fundamental physical theory.

A few years after arriving in Pittsburgh, I wrote a paper on one of these conceptual problems that resulted in an invitation for me to pursue a model for the neutron at the Institute of Theoretical Physics of the University of Paris. The Institute was directed by Louis de Broglie, who had won the Nobel Prize for predicting the wave nature of the electron in his thesis in 1923, the same year that Millikan had won the Nobel Prize for showing that the electron was a point-like particle with a unique, quantized charge.

A brief discussion of the themes behind my paper may elucidate some of the central issues in fundamental particle theory in the post-War years. For more than twenty years, a certain thought-experiment

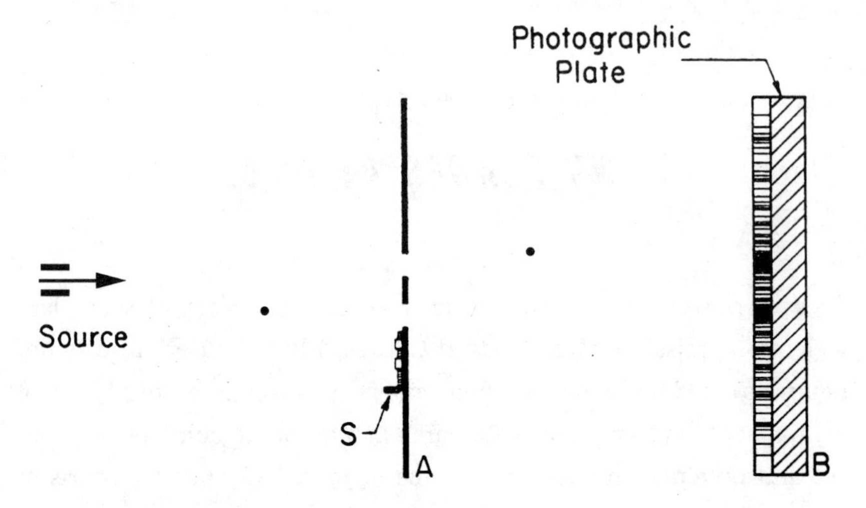

Figure 7.1 Schematic diagram of idealized double-slit interference experiment with point-like particles.

had been used by Bohr and other physicists of the Copenhagen School to argue that our ordinary causal space-time description of physical phenomena breaks down on the atomic scale. That was in opposition to Einstein's view that such a radical departure from our normal mode of describing natural phenomena would turn out to be unnecessary. The linchpin of the Copenhagen School's argument was known as the double-slit interference experiment, and was often cited in textbooks of quantum mechanics as demonstrating the breakdown of classical physics.

The argument is summarized by Figure 7.1. A stream of particles is fired at screen *A* containing two parallel slits. The particles are recorded on screen *B*, which holds a photographic plate.

When both slits are open, an interference pattern is produced by electrons, similar to one seen with light waves. This pattern arises as a result of the different distances that the light waves travel from the slits to various points on the photographic plate. Wherever crests of the waves emanating from the two slits occur in the same place, after developing the film one finds a higher number of dark silver grains in the photographic emulsion than in areas where the crest of one wave arrives together with a trough in the wave coming from the other slit, since the effect of the two waves tends to cancel each other. Closing the shutter *S* for one of the slits alters the pattern. Now it appears as one seen with light waves produced with a single slit, slightly fuzzy at the edges, since light waves can travel around the edges of obstacles to some degree, just as ocean waves travel around the end of a sea-wall.

Incredibly, when particles are fired at the screen *one at a time* with both slits open, even when matter particles such as electrons are used, after enough electrons have accumulated in the photographic emulsion, the double-slit patterns—and not two single-slit patterns—are still observed. This makes it impossible to explain the double-slit pattern formation in terms of an effect of electrons passing through one slit on those passing through the other slit, so that only the following two equally bizarre explanations remain, as discussed by the philosopher of science Henry Margenau in his 1950 book *The Nature of Physical Reality*. One either has to assume that the electrons possess sufficient intelligence to "know" whether the other slit through which they are *not* passing is open or closed, or that they somehow have "an inherent disposition" to go through one slit rather than the other.

On the basis of this dilemma, particularly puzzling in the case of electrons envisioned as possessing a point-like character with a highly localized charge or mass, Bohr and his followers concluded that the concept of a continuously varying position in space and time during the process of scattering has no physical meaning at all. They argued that one cannot follow the path of an individual electron, so that only the final result of the experiment can be described, and then only in statistical terms giving the probability of an electron arriving at a certain point on the photographic plate. Therefore, causality and the usual space-time geometric description of natural phenomena necessarily fail on the atomic scale. As Bohr put it in a 1955 lecture at a meeting of the Royal Danish Academy of Sciences, published in his book *Atomic Physics and Human Knowledge*, "in proper quantum processes, we meet regularities which are completely foreign to the mechanical conception of nature and which defy pictorial deterministic description."

However, on the basis of the purely electromagnetic description of electrons that I had discussed with Einstein, it seemed to me that there was a basic fallacy in this argument. The account of the double-slit experiment tacitly assumed that *an electron can pass through only one or the other of the two slits and not through both at the same time.* It meant, in effect, that electrons are taken to be describable only as either infinitely small points or as small "hard balls" of sharply defined "size," similar in all essential respects to Margenau's analogy of bullets fired through slits in an impenetrable steel wall. But in a purely electromagnetic concept of an electron, it cannot be point-like in nature because a charge compressed into an infinitely small volume has an infinitely large mass. Nor would there be any hard "ponderable" matter of sharply defined, finite size at its center. Instead, in a classical electromagnetic model with the charge distributed over a finite source for the field, all the mass resides in the electromagnetic field in the surrounding space, with no energy or mass at the exact center at all. Thus, for an electron accelerated to some 50,000 volts and moving at nearly half the speed of light, the electron's field and the source from which it originates would be highly flattened according to the formula derived by Lorentz and, on the basis of different assumptions, by Einstein in his theory of special relativity. As a result, the electron's field energy, and therefore its mass according to the relation $E = mc^2$, would be compressed into a disk of growing size as its velocity approached that of light and its electric field at right angle to its direction of motion would grow stronger. When I calculated the distance over which the field of such a relativistic electron could give a certain minimum possible amount of energy to the atoms in a metal screen comparable to that found in its normal vibrations, it turned out to be of the order of the space occupied by some 250,000 atoms in a row—some twenty times larger than the distance between the two slits used in the actual experiments. Thus, an electron regarded as an extended field structure similar to that of a photon can in fact "go through both slits at a time," as shown in Figure 7.2 where the experiment is shown schematically. It is actually carried out in the laboratory with one electron at a time.

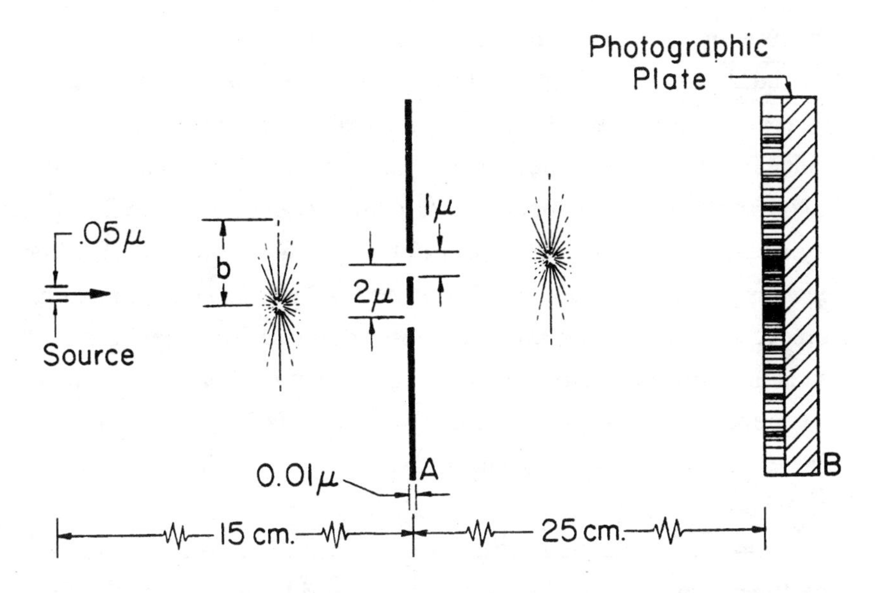

Figure 7.2 Schematic diagram of double-slit interference experiment as actually carried out with single electrons by J. Faget and C. Fert (Cahiers de Physique, **83**, 285–292 (1957). (1 μ = One millionth of a meter)

Accordingly, using the purely electromagnetic description of mass or energy distributed in the space around a charge as proposed by Lorentz in the early 1900s, there is no conceptual difficulty in visualizing a matter particle such as an electron moving in a normal, classical, detailed geometric or "mechanical" manner along a continuous trajectory. The center of an electron can go through one of the open slits while, at the same time, a part of its field goes through the other slit. The conceptual dilemma only arises if one uses an inappropriate model for the electron as a point-like or a hard "billiard-ball-like" entity falling on a slit system in a solid wall, as Margenau proposed. On the atomic scale, the "solid wall" is really mainly empty space.

As I learned from the literature, the same situation is found in all other types of electron interference or diffraction experiments that can actually be carried out, using thin wires or other arrangements. In all cases, the size of the region over which the electron's field and thus its mass-energy is distributed—and through which it interacts with other systems composed of electrons and protons—is necessarily comparable with the distance between scattering centers. If this condition is not met, it becomes impossible to detect the pattern of light and dark interference fringes, because they are too faint and too closely spaced. Thus, the conceptual dilemma that appears to force a renunciation of the ordinary causal space-time description of phenomena only arises when electrons and protons are seen as other than purely electromagnetic. Moreover, as I was to fully appreciate only many decades later, the electron is not only associated with an electromagnetic field that stretches out to appreciable distances, but also with a gravitational field that is not affected by any intervening barriers, such as the portion of the metal screen between the two slits.

As Einstein wrote me in responding to my essay on the concept of time, the concepts we use in describing the world of phenomena such as those of space and time discussed by Kant

> cannot be deduced by logical processes from experience as the empiricists had in part believed and still believe. This is the case not only for space, time and causality, but for all concepts, for all that which is communicable by means of language. But it is not true that there exist *a priori* certain *specific* concepts in the sense that thinking without them is impossible. The only thing that is true is that thinking is an operation with concepts, so that thinking is impossible without them. But as to just what these concepts should be, one cannot know anything *a priori.*

Clearly, Einstein's remarks apply to the double-slit experiment, where the choice of a concept for matter other than the classical Kantian or Newtonian ones of point-like or finite-sized "ponderable" spheres has to be used. As had emerged in my conversation with Einstein, the problem was to find a concept for a stable, purely electromagnetic electron that would relate it intimately to a stable photon with wave-like properties—and yet preserve the point-like interaction of both with other particles, as observed in numerous experiments.

The stability of an entity composed of nothing but an electric charge posed a problem. At the time I wrote the paper on the double-slit interference problem, I had not tried to explain the stability of the electron, a problem that had existed since its discovery at the end of the nineteenth century. I decided to simply accept a stable source of an electric field, or fundamental charge, for the purposes of understanding the heavier particles, rather than attempt to explain why such a charge did not break up into smaller pieces due to the repulsion between like charges. In any case, such an attempt would be based on the questionable assumption that an ultimate entity can be treated exactly the same as a macroscopic body composed of many such entities, an assumption already seriously questioned by Lorentz in 1906. It involved the empirically-unfounded belief that "sub-elementary charges" exist, distributed in some manner inside the electron, exerting forces on each other that need to be balanced. But in over half a century of experiments at increasingly higher energies, no evidence showed the electron could be broken up into smaller charges.

I argued that a finite size for the region from which the field of the electron originated, necessitated by its finite mass, did not mean still smaller charges exist in nature. That would confuse a theoretical conceptual *possibility* with a physical *actuality*. This reasoning appeared to me on a par with the argument that some particles might *in principle* move at velocities greater than the observed speed of light, a statement that might be made if mass was other than electromagnetic in origin, even though photons and all particles like electrons were *in fact* never found to exceed the speed of light under any experimental conditions. This was the fundamental assumption that Einstein used to develop a maximum velocity for the motion of particles or the transmission of signals in his Special Theory of Relativity, and it has never been found to fail in nearly a century of physical experiments.

All the evidence suggested that the electron can be created or destroyed only as a complete unit, together with its oppositely charged anti-particle, the positron, forming two gamma rays or photons of purely electromagnetic nature, so that in this sense these two entities appeared to be the only truly indivisible, structureless and stable elementary particles. The high-energy electron-proton scattering experiments of Robert Hofstadter at Stanford University strengthened this assumption in the mid-1950s: the proton appeared to have a complex structure.

The work of Emilio Segré and Owen Chamberlain at the Berkeley Laboratory of the University of California in 1955, for which they won the Nobel Prize, confirmed the proton's nature. Segré and Chamberlain established that various mesons—that is, smaller particles—result from the annihilation of protons and anti-protons. By contrast, the positron and the electron annihilate into radiation: two purely electromagnetic gamma rays.

When a piece of material like lead is bombarded with high energy protons, the proton being struck gets internally excited but does not disintegrate. It absorbs the energy and then gets rid of it by creating a proton/anti-proton pair. The anti-proton, after very short travel in ordinary matter, slows down, then gets captured by a proton, which it merges with and, in doing so, destroys itself and the proton. Charge is thereby conserved. When the negative charge of the anti-proton disappears, so does the positive charge of the proton. In the course of this annihilation, a minor explosion results which can be tracked in the bubble chamber—from which we know the anti-proton and the proton are composed of mesons, because we see them flying out.

Accepting the electron and the positron as stable, extended sources of electromagnetic field energy also explained why an electron, described as a localized "wave-packet" by de Broglie and Schrödinger, does not immediately "run apart." This early objection to a real wave-packet interpretation of a particle seemed to me equivalent to the question of why the electron does not blow up due to the mutual repulsion of electrostatic charges assumed to compose it. That a theoretical wave-packet, constructed mathematically from an infinite series of harmonic waves, spreads out rapidly cannot be accepted as a valid argument against the existence of a stable electron. Instead, I argued that it meant our existing concepts and mathematical treatments of the fundamental charges needed to be revised.

If a fundamental charge is nothing but the source of a field, the conflict between the atomistic and the continuous view of matter resolves itself since "matter" and "field" become identified with each other. Neither "particles" nor "fields" can exist separately in this view, nor need either concept be conceded to have a more fundamental character than the other. In fact, as I argued in my conversation with Einstein, the field of a charge moving close to the speed of light is indistinguishable in its interaction with matter from that of a photon.

Fermi and others used this idea to treat the interaction of electrons as being produced by the emission and absorption of "virtual photons." (A virtual photon, or the contracted field of an electron, is a wave pulse, and behaves as if it were a photon.) Since, in an electromagnetic theory of matter, the electric fields originating and ending in these sources define physical space, the field lines themselves could also be regarded as the carrier of the electromagnetic waves generated by the acceleration or deceleration of the sources. They behave like waves moving along a violin string. This was a view that had been advocated by Faraday in his *Thoughts on Ray Vibrations* more than a hundred years ago.

It so happened that at the time I had written a draft of this paper in the summer of 1956, a French engineer was working at the Westinghouse Research Laboratory. In the course of a conversation, I mentioned my paper on the double-slit experiment, and told him that I had been interested for a long time in the problem posed by the wave-nature of the electron arrived at by de Broglie. Thereupon he told me that he knew de Broglie, and offered to bring him a copy of my paper on his return to Paris in the next few weeks. To my great surprise, a few months later I received a letter from de Broglie. He invited me to spend a year at the Institute of Theoretical Physics of the University of Paris, working on my ideas on the conceptual problems of the electron and the other particles of physical theory.

Much as meeting de Broglie and working full-time on the fundamental problems of physical theory appealed to me, it was difficult for me to accept his invitation both for personal and professional reasons. Following the traumatic experience of seeing a beautiful and apparently healthy newborn baby slowly succumb from a subtle genetic disorder that would prevent us from having any children, my twelve-year marriage was ending in divorce. At the laboratory, I had just been joined by two young German physicists, Helmut Kanter and Gerhard Goetze, who had learned of my work in secondary electron emission and were eager to pursue the interesting applications to particle physics and electronic image intensification in medicine and astronomy. My responsibilities had greatly increased. Moreover, there was no formal policy of sabbatical leaves, such as exist in the academic world, that would allow me to undertake a year off for research.

However, the new director of the laboratory, Clarence Zener, wanted to encourage intellectual exploration, and decided to institute a program of sabbatical leave when I told him of de Broglie's invitation. This decision allowed me to plan a stay in Paris beginning in the fall of the following year while the experimental work in electron physics would be continued by my new colleagues. And so I wrote a letter to de Broglie indicating that I would accept his invitation in September of 1957.

In January of the next year, I met a young woman by the name of Marilyn Seiner as a result of a blind date arranged by one of my friends at the laboratory. She was a graduate of Carnegie-Mellon University, and at the time she was working in an advertising agency in Pittsburgh, the city where she was born and had lived all her life. Early in our acquaintance I mentioned to her that I was planning to spend a year in Paris beginning in September, and so neither one of us expected that our relationship would last. However, by mid-summer we had gotten to enjoy each other's company so much that when I asked her whether she would be willing to come to Paris with me, she happily agreed to arrange for the wedding to take place in the ten weeks before my scheduled leave was to begin. The ceremony attended by her large family and many friends took place the weekend before our departure, so that the stay in France became a long honeymoon that turned out to be the beginning of a wonderful new life, starting with a five-day sail across the Atlantic on the "Flandres" that left New York for Le Havre on my birthday.

On our arrival in Paris, we were greeted by my brother and his wife who were spending their first year of marriage in Europe, and who took us to a small hotel on the Left Bank not far from the Sorbonne and the Institute of Theoretical Physics. Unfortunately, I had come down with a severe case of influenza on the last day of our trip, and it was a few weeks before I was able to meet de Broglie.

As I walked through the narrow streets of the historic Latin Quarter to meet de Broglie and saw the sign for the Rue Pierre Curie, I was reminded that it was here in Paris that radium was discovered. That development set in motion the events that led to the discovery of the proton and neutron, whose structure was still a mystery I was struggling to understand, half a century after they had first been found.

As I entered the courtyard of the Institute's large red-brick building, I thought of the man it was named after, and who had taught at the University of Paris at the turn of the century when the Curies carried out their research: the renowned mathematician Henri Poincaré. Poincaré had suggested that the stability of the newly discovered electron might be explained by a locally strong attractive force of unknown character. Poincaré also shared Lorentz's view of the electromagnetic nature of mass. As early as 1899, he had theorized that optical phenomena depend only on the *relative* motions of material bodies, and that no velocity can exceed that of light, and that perhaps the ether did not exist at all. Einstein developed these ideas more fully and became known for them in the decades after Poincaré's death in 1912 at age 58. While I was a graduate student, I had been enormously affected by reading Poincaré's book *The Foundations of Science*, in which he called for a re-examination of all our existing physical concepts, to consider the possibility of a non-Euclidean geometry for our actual space. These ideas influenced the young Einstein and developed into his General Theory of Relativity to explain gravity a few years later.

There could not have been a more appropriate place for me to wrestle with the conceptual problems of the electron and the other particles of matter than the Institute named after Poincaré and now headed by de Broglie.

CHAPTER 8

LOUIS DE BROGLIE

THE PARADOX OF LIGHT—its particle nature along with its clear wave properties—led Einstein to ask me whether I had any ideas what a photon might look like. This question was intimately related to the one which caused de Broglie to attempt to visualize the nature of the matter waves for particles that he had postulated back in the early 1920s. In the course of my stay in Paris, I would learn of the different efforts to solve this puzzle that he had made over the years.

The most recent was in cooperation with his young associate Jean-Pierre Vigier and the American theoretical physicist David Bohm, an expatriate living at that time in Brazil, where Vigier had first gone to meet him. The problem posed by the wave properties of the electron had brought me to Paris: but the most important result of my fourteen month stay there was the discovery of a way to calculate the forces between two charges in relative motion with respect to each other, published in 1905 by Einstein in his groundbreaking paper on relativity and completely overshadowed by his revolutionary attack on Newton's absolute space and time.

Einstein died the year before de Broglie's invitation to me arrived. His nearly forgotten suggestion was to provide the key to the nature of mesons and the structure of the proton and neutron. Decades later, his results for the force between two highly relativistically-moving charges and their increase in mass with growing relative velocity would provide the basis for understanding the mass and density of the Lemaître atom, as well as the origin of the cosmological structures in an expanding universe. But only after decades of work by many researchers—often on questions that seemed unrelated to the nature of matter waves and the stability of the electron—did these developments finally lead to a unified concept for the nature of photons and electrons. The trick was to both explain their stability and provide a resolution of the wave-particle duality problem, with which de Broglie had been wrestling for more than three decades.

When I met de Broglie in his office at the Institute of Theoretical Physics, he was in his middle sixties, a few years younger than Einstein was when I had visited him ten years earlier. They both believed that one must find ways to describe natural events, even on the atomic scale, in the framework of space-time that permitted the development of clearly visualizable models, and they had in common a preference to solitary work. But they differed radically in their appearance, their personality and their background.

De Broglie was a man of slender build, with carefully groomed dark hair, very formally dressed and equally formal in his soft-spoken, rather diffident manner that seemed consistent with his family's long aristocratic background. After welcoming me to his Institute he introduced me to Vigier, who spoke English fluently. De Broglie preferred to speak French, which made conversation difficult since my high school French was very rusty. Thus, most of the time I discussed physics with Vigier, and what I learned of de Broglie's ideas was mainly through his many articles and books, some of which I had read in their English translation.

Prince Louis de Broglie was born in 1892, the younger of two sons of a noble family that traced its ancestry to the tenth century and whose members had served the monarchy in military, diplomatic and political positions for centuries. Early in his life, de Broglie was attracted to the study of philosophy and then history, like his grandfather, who had written a book about the medieval church. His father served as deputy in the French parliament, and young Louis, initially educated privately by a priest, was expected to become a statesman. However, when he was sixteen, his father died and his older brother Maurice, who had revolted against family tradition and become a physicist, took over the responsibility for his younger brother's education, who also revolted against family tradition and became increasingly interested in the history and philosophy of science. In particular, de Broglie was enormously impressed by the books of the philosopher Henri Bergson and the mathematician Henri Poincaré, the former paving the way for the revolutionary ideas of quantum theory and the latter developing the equally revolutionary theory of relativity with its rejection of Newton's absolute time and reference frame.

De Broglie disagreed with Poincaré in one important respect. Poincaré believed that a great number of possible theories could explain given phenomena equally well. De Broglie felt that there could be one particular theory that conforms more to the underlying physical reality than all others, and therefore that it is "more apt to reveal the hidden harmonies." He was intrigued by Bergson's ideas, according to which the trend in physical theory pointed to a dissolution of the atom into wave-like movements or lines of force that would permit a return to a universal continuity. Bergson referred to matter as vortex rings whirling in a continuous medium as advocated by Lord Kelvin, and the vision of Faraday, for whom particles of matter dissolved into lines of force extending throughout physical space that were the carriers of electromagnetic waves.

De Broglie obtained a degree in science in 1913 after having obtained a degree in history and philosophy, shortly after which the outbreak of war forced him to interrupt his studies. He entered the French military and worked in the wireless service, where he became involved in a project to utilize the Eiffel Tower for military communication using radio waves.

As described by his brother Maurice in *Louis de Broglie: Physicien et penseur,* after leaving the military in 1919, de Broglie worked in his brother's laboratory studying the wave properties of X-rays, which led him "to reflect profoundly on the necessity for always associating the point of view of waves with that of corpuscles." His experience with transmitting messages in the form of pulses of radio-waves, together with his effort to understand the particle nature of electromagnetic waves, or photons, discussed in Einstein's photoelectric theory, apparently played a role in his asking himself whether matter particles might not also possess wave properties. His study of the history of the theories of light as waves and the alternative description of light as a form of corpuscular motion going back to Descartes, Fermat, Huygens, Newton and Maupertuis led him to the novel idea that waves were associated with matter particles which guided their motion. Similarly, electromagnetic waves seemed to guide the motion of the corpuscles of light, which Einstein needed to explain the photoelectric effect. Both light waves and particles obeyed similar laws, according to which they would always move along paths that required the least time (as originally found by Fermat), or the least "action" (as discovered subsequently by Maupertuis and Hamilton), where action was the product of momentum—which is mass times velocity, times the distance traveled.

This similarity between particles of light and matter was reinforced in 1922 by the discovery by the American physicist Arthur Compton that energetic X-rays were scattered by loosely bound electrons, exactly as if they were matter particles. Some of the momentum of the X-ray was lost by the X-ray photon and transferred to the electron, with the X-ray photon being scattered in a certain direction like a form of "needle radiation," rather than being diffused in waves. This scattering was exactly what would be expected if the photon and the electron were particle-like in their dynamic properties.

By September of 1923, de Broglie had worked out the mathematics of his "matter-waves," and submitted the first of a series of articles to the Proceedings of the French Academy of Sciences. What de Broglie had found was that the momentum should be inversely related to the length of a wave associated with a particle. De Broglie was thus able to show that for electrons moving very close to the speed of light, matter waves would have the same wavelength as those of light or X-rays, and that for matter particles moving at lower velocities the wavelength or "size" associated with the particle would increase, while an associated frequency would decrease. He predicted that, based on the symmetry of the laws that governed the motions and collisions of photons and electrons, matter particles would show the same kind of diffraction and interference effects seen with light waves and X-rays.

By coincidence, without de Broglie's knowledge, this prediction had already been verified by experiments carried out in the United States in early 1923 by C. J. Davisson and C. H. Cunsman at the Bell Telephone Laboratories. Davisson and Cunsman found unexplained increases in the number of reflected electrons from a platinum target in certain directions.[1] In 1928, Max Born, professor of theoretical physics at the University of Göttingen, heard of these experiments and suggested to one of his students, Walter Elsasser, that he should investigate whether these patterns of reflected electrons could in some way be explained by de Broglie's ideas, according to which electrons of a certain momentum or energy should be diffracted like X-rays by the planes of atoms in the platinum target. Elsasser's confirmation of this possibility was the first evidence in support of de Broglie's thesis. It set in motion a whole series of important developments that revolutionized physics.

One of the most important of these was de Broglie's explanation of why there were only certain discrete orbits allowed for electrons in Bohr's model of the hydrogen atom, the question that had fascinated me as a young boy. Bohr's quantum condition was that electrons had to have an angular momentum in their orbits around the proton that was a multiple of a certain minimum value given by Planck's Constant h divided by 2π, symbolized by $\hbar$. The angular momentum is given by the product of the radius of the orbit and the electron's mass and velocity.

The orbits that can exist are those where the circumference can accommodate exactly a whole number of electron wavelengths. For the lowest or ground-state circular orbit, de Broglie showed that the electron wave must return exactly to the same place after completing one revolution in its orbit, or that the wave representing the electron must fit exactly into the orbit, just like the lowest harmonic in a violin string "fits" exactly into the length of the string. Thus, the description of the electron as an extended, pulse-like entity occupying a certain space explained the mysterious rule for the particles of Bohr's atom in a wholly new way: in terms of modes of vibration of the electric fields constituting the atoms.

The year following Elsasser's conclusion that experiments confirmed the reality of de Broglie's matter waves, the theoretical physicist Erwin Schrödinger developed a detailed wave-description of the hydrogen atom based on an equation that described the amplitude of the matter wave in the atom and thus the position of the electron, in much the same way as Maxwell's equations for electromagnetic waves gave information on the position of light quanta. For Schrödinger, the matter waves were as real as the electromagnetic waves of Maxwell, though he was unable to specify exactly the physical nature of the matter waves. On the other hand, Born suggested that the amplitude of the matter wave (known as the "Psi function") was simply a measure of the probability of finding an electron at that point in space, so that it was not a physically real entity, and instead a mere "probability wave" of purely formal or mathematical character. Born's view was consistent with the ideas of his young assistant Werner Heisenberg. Heisenberg had published a paper on a purely algebraic approach to the hydrogen atom shortly before Schrödinger had published his wave-mechanical model for the hydrogen atom. It involved a mathematical method using rows and columns of quantities related to the transitions from one allowed state of the atom to another, called matrix mechanics, in which no attempt was made to arrive at a visualizable picture either of electron orbits or the matter waves of de Broglie.

In Heisenberg's matrix mechanics, the only legitimate questions allowed to be asked about the behavior of an atomic system involved "observables," such as the mathematical probabilities of the emission of quanta of light of different frequencies. Detailed, deterministic space-time models of atomic systems of the type used in classical physics were regarded by Heisenberg as unnecessary and inconsistent with a proper operational approach he felt Einstein had used in formulating the Special Theory Relativity.

In 1927, Heisenberg's aversion to geometric or dynamic models led him to write on the inherent impossibility of formulating precise models on the microscopic scale. This was due, he wrote, to the limitation set by the existence of a minimum quantity of "action" in quantum physics, namely the amount given by Planck's Constant h. In his paper on the so-called "Principle of Uncertainty" Heisenberg argued that it will always be impossible to simultaneously measure the position and the momentum of a particle with perfect accuracy.[2] This analysis supported Heisenberg's deeply held view that "deterministic" or "causal" physical models with definite orbits of electrons are really a myth, that only probabilistic models such as his matrix mechanical approach would be useful in theorizing on the still smaller level of nuclear particles, and that therefore quantum mechanics in the statistical version that he and Born had arrived at was in its complete and final form.

This was the beginning of a long and often bitter controversy over which approach to theoretical physics was better for arriving at a more complete, unified understanding of the nature of matter and light, a controversy that had not been resolved thirty years later when I met de Broglie. Originally, when de Broglie first published his ideas on matter waves, he believed that they represented a real physical wave which, like its electromagnetic field, was somehow associated with an electron. Two years later, he arrived at the idea of "the theory of double solutions," according to which Schrödinger's equation allows two different solutions: one that is continuous, where the wave function has a statistical significance, and another that has a solution with a singularity, where the singularity constitutes the physical point-particle under consideration. This approach preserved the classical idea of the particle as a source of a field, guided by an extended wave that subjects the particle to diffraction and interference which may be produced by far-off obstacles. In an effort to achieve a wave-particle synthesis of this kind, but without a singularity at the center of the electron, I wrote my paper on the double-slit interference experiment that appealed to de Broglie in 1956. He had abandoned such an approach in the late 1920s, under pressure of the predominant purely probabilistic interpretation of quantum theory with its many practical successes in atomic physics.

Under the influence of David Bohm, de Broglie had returned to a realistic space-time or "causal" view of physical phenomena by the time he invited me to his Institute. Early in the 1950s, a young mathematician by the name of Jean-Pierre Vigier came to the Sorbonne to work with de Broglie. In 1952, Vigier became aware of recent papers by Bohm, at that time an assistant professor of theoretical physics at Princeton and the author of a textbook on quantum theory that largely followed the views of the prevailing Born-Heisenberg interpretation. In sharp contrast, the papers that Bohm published since writing his textbook questioned the probabilistic interpretation of the wave-function and the impossibility of arriving at a causal description of microscopic phenomena. Bohm believed there were "hidden variables" on a deeper level that would eventually allow a deterministic description of phenomena on the microscopic scale.

As described by Bohm in his 1957 book, *Causality and Chance in Modern Physics*, after he arrived at Princeton in the fall of 1946, he had a series of conversations with Einstein. Bohm became increasingly convinced, as was I, that progress in understanding the newly discovered particles and the nature of the wave-particle duality of matter would require detailed space-time models and variables on a deeper level than those permitted by Heisenberg and Born's purely probabilistic approach. As Bohm put it:

> . . . Let us recall that one of the principal problems now faced in this domain is that of treating the structure of an "elementary" particle, and of discovering what kinds of motions are taking place within this structure—motions that would help to explain, perhaps, the "creation" and "destruction" of various kinds of particles, and their transformation into each other. In the usual interpretation of the quantum theory, it is extraordinarily difficult to consider this problem. For the insistence that one is not allowed to conceive of what is happening at this level means that one is restricted to making blind mathematical manipulations with the hope that somehow one of these manipulations will lead to a new and correct theory.

As de Broglie indicated in the foreword to Bohm's book, Bohm took up ideas from de Broglie's 1927 article on the "theory of double solutions" which represented a causal interpretation of wave mechanics.

> . . . Commenting and enlarging upon them in a most interesting way he has successfully developed important arguments in favor of a causal reinterpretation of quantum physics. Professor Bohm's paper has led me to take my old concepts up again, and with my young colleagues at the Institute Henri Poincaré, and in particular M. Jean-Pierre Vigier we have been able to obtain certain encouraging results.

De Broglie also mentioned in his foreword that Vigier began to collaborate with Bohm in interpreting the wave function. De Broglie wrote: "It seems desirable that in the next few years efforts should continue to be made in this direction. One can, it seems to me, hope that these efforts will be fruitful and will help to rescue quantum physics from the *cul-de-sac* where it is at the moment."

In conclusion, de Broglie added a sentence that summarizes why I felt so fortunate in being able to discuss my work with him and Vigier at that time: "Convinced that theoretical physics has always led, and will always lead, to the discovery of deeper and deeper levels of the physical world, and that this process will continue without any limit, he [Bohm] has concluded that quantum physics has no right to consider its present concepts definitive, and that it cannot stop researchers imagining deeper domains of reality than those which it has already explored."

It was Einstein's lonely persistence in questioning the completeness of quantum theory and its purely probabilistic interpretation of matter-waves, despite widespread opposition, that led to de Broglie's change of view. It explained why I had been invited to work on an effort to find new concepts needed to develop detailed space-time models for nuclear particles.

As I learned from Vigier, Bohm and a former fellow graduate student of his studying with Oppenheimer at Berkeley, Martin Weinstein, had published a paper in 1948 in which they used a classical model of the electron with an extended, finite source capable of internal excitation in an effort to explain the pi-meson mass. They hoped, as did I, that this approach would also overcome the problem of the infinities encountered in dealing with point-charges in quantum theory.[3]

While Vigier continued to work with Bohm to develop a model of the electron involving relativistically moving fluid masses, I decided to pursue the model of the neutron as composed of an electron and a proton that I had begun to develop at Cornell. Learning of Bohm and Vigier's use of relativistic hydrodynamics, or the theory of fluids at very high velocities, motivated me more strongly than ever to work on a neutron model understandable in terms of purely electromagnetic entities of finite size.

Bohm and Vigier also used the ideas of vortices or whirlpools to describe the electron and its spin. Over the years, I had become increasingly convinced that only if the neutron and all other particles and their interactions could be described in electromagnetic terms could there be a unified view of photons and electrons, a hope that de Broglie and Einstein kept alive despite formidable opposition. I felt that only if a link existed between electromagnetic and nuclear forces could one fully accept the concept of the ether as an ideal fluid capable of sustaining rotary or vortex motions. Only if all particles were ultimately constituted of nothing but sources of electromagnetic fields, so that their mass could be understood as being of purely electromagnetic field origin, would it be possible to explain the failure to detect the ether in the Michelson-Morley type of experiment. That failure was then ascribable to a contraction of measuring rods and a slowing-down of motions in a moving reference frame such as that of the Earth, as had been worked out by Lorentz.

In *Matter and Light,* published by de Broglie in 1937, he speculated that photons might consist of a neutrino and an anti-neutrino. That agreed with my feeling that the recently discovered neutrino, emitted together with an electron from a neutron in beta-decay, might also eventually be explainable in electromagnetic terms. De Broglie's other conjecture that the electron and positron might ultimately be understood as somehow composing a photon, supported my belief that the particles of matter and light would turn out to be closely-related manifestations of a single underlying ideal substrate, whose local curvature in the neighborhood of matter also determined the action of gravity—a step towards the unification of all forces.

In the case of the neutron model, one of the problems that remained was how close an electron could approach the proton in the capture process discovered by Rutherford's colleague, the mathematician C. G. Darwin. At the time, it seemed to me that this would be determined by the finite sizes of the two particles. Only recently had an accelerator for electrons been completed that had enough energy, and thus momentum, to produce electrons of sufficiently short de Broglie wavelength to begin to disclose the size and structure of the proton. This machine, a large linear accelerator, had been built at Stanford University in the physics department headed by Robert Hofstadter. I had met Hofstadter through my development of thin secondary electron emitting foils at the Westinghouse Research Laboratories. Hofstadter had decided to use the foils in monitoring the electron beam, and so I followed his work with great personal interest.

This, the most powerful electron accelerator then in existence, was eventually able to accelerate electrons to an energy of a billion or 10^9 volts in a 150-foot-long evacuated tube before they struck their targets, using powerful radio-waves in a series of steps to reach the final energy. At the maximum energy achieved, the electrons had a very small de Broglie wavelength. The shorter the wavelength, the finer the detail that can be resolved by a microscope, and Hofstadter's accelerator was in effect a very powerful microscope. The ability of Hofstadter's accelerator to resolve detail was about a third of the theoretical size of a classical electron with its charge distributed over a radius of 1.4×10^{-13} centimeters, or about half of ten trillionths of an inch. This was also roughly 100,000 times smaller than the orbit of the electron in the hydrogen atom, and nearly equal to the size of the proton as earlier determined by Rutherford. However, in the first experiments in the early 1950s, the available energy was only about one-fifth the billion volts eventually reached, so the initial experiments were used to determine the positive charge distribution of the nuclei for elements heavier than hydrogen.

By 1954, the energy attained in the accelerator had become great enough to begin studies of the charge distribution of the proton. In 1956 the first results were published, showing that the proton's charge was not a point-like entity in the center. The data showed instead that the proton's charge was distributed over a surprisingly large region, with an average radius about half that of the classical electron, while its outer size was twice this size, or about equal to that of the classical "shell" electron. This immediately implied that the mass of the proton could not simply be explained by the energy in the electromagnetic field of a very small single positive charge, scaled down by a factor of some 1,836 times, the ratio of the proton to the electron mass. Instead, the experimental results as interpreted by Hofstadter indicated that most of the mass must be due to a cloud of mesons, and that therefore the proton had a complex structure, unlike the electron.

Vigier and de Broglie were very interested in these new results. I decided to write a short paper on the implications of Hofstadter's experiments for the hypothesis that the electron might have an extended source for its electric field, which Vigier translated into French and de Broglie submitted for publication in the proceedings of the French Academy of Sciences. I followed this up with a second paper in March of 1958, in which I discussed the problem of the nature of the electron as a stable, extended source of a field along the lines of the double-slit article, which de Broglie had also submitted for publication in the Academy's proceedings.

Hofstadter's experiments did not conflict with my original idea that an electron could circulate in an extremely small orbit, roughly of the classical shell radius in size, with an angular momentum one half the ordinary Bohr orbits of hydrogen. Its own spin angular momentum would thereby be canceled, explaining why the neutron had the same absolute value of spin as the proton. Moreover, for capture of an electron at such a distance from the center of the proton, the momentum, and thus the energy, were in rough agreement with the maximum energy of 780,000 volts with which electrons had been found to emerge from neutrons in their decay, so such a composite model was not ruled out by the new data of Hofstadter.

In fact, the first data on the charge distribution of the neutron just then being gathered at Stanford clearly showed that it also had a complex structure. There appeared to be a circulating negative charge producing a magnetic field at about the same average radius where the positive charge was found. My calculations of the relativistic mass that an electron would acquire in the course of spiraling in towards the proton would be about that of the known mass of the pi-meson, in agreement with Yukawa's prediction for his heavy "nuclear electron." The de Broglie wavelength associated with the large relativistic mass could be fitted into the small orbit, just as in the case of the much larger hydrogen orbits.

But perhaps the most important result was that the strength of the force between the relativistic electron and the proton turned out to be on the order of the known strength of nuclear forces. Lorentz had shown that if all forces, including those acting between protons and neutrons in the nucleus, were due to the action of electrical charges—as was found for the forces between the electrons and protons in atoms and for the forces between neighboring atoms—measuring rods moving through the ether would contract and motions such as those in clocks would slow down exactly so as to make it impossible to detect motions relative to the ether. Only if it were to turn out that all forces in nature are explicable in terms of the forces acting between charges could one assume that the ether exists. Therefore, if the forces between neutrons and protons could be explained in terms of electrons moving around them at velocities close to the speed of light when their fields become highly flattened and thus very strong, it would be possible to return to the old ideas of the ether as an ideal fluid and matter as vortices.

Nor was the recent definite proof of the existence of neutrinos being emitted from a neutron along with an electron a serious problem for my model of the neutron or the existence of the ether, since the neutrino appeared to be a special form of electromagnetic radiation. In fact, de Broglie had speculated that except for its smaller spin and greatly reduced interaction with ordinary matter, the neutrino could be a photon with half the spin, having no rest-mass, just as the photon.

Finally, all the new evidence for a complex structure of the neutron and proton, combined with the rapidly mounting evidence for excited states of the proton being produced in high energy machines all over the world, appeared to justify Bohm's insistence that there must be structure found at deeper levels that required further variables. The new evidence supported the idea that the purely probabilistic Copenhagen interpretation of quantum theory was in fact inadequate for dealing with the structure of nuclear particles.

All these hopeful developments were further supported by a discovery I made in the library of de Broglie's Institute a few months later. In a new book on electrodynamics written by a French physicist by the name of Arzélies, I found a simple calculation of the forces between two very rapidly or relativistically moving charges that I had not seen before. It turned out to be based on a suggestion made by Einstein in his famous 1905 paper in which he advanced his Special Theory of Relativity. Einstein had written that, to avoid an asymmetry in the force calculation for charges in relative motion with respect to each other, one has to calculate it as measured by an observer at rest with respect to one or the other charges. As Arzélies pointed out, this meant a consistent answer is unobtainable if the force is calculated by an observer in another reference frame, such as the laboratory where the center of mass of the particles might be at rest. For the particular case of an electron moving around a proton as in my neutron model, this meant that the force could be described as a strengthened electrostatic force.[4]

The Lorentz factor now came into play. It is a pure number that increases from the value one at zero relative velocity of the two charges to an infinitely large value as the speed of light is approached, in effect making the finite speed of light play the role of an infinite velocity in the physical world. It means that the strength of the electric field measured by the number of lines of force per unit area, or their density, produced by a moving charge becomes increasingly greater as the speed of a passing charge approaches that of light, some 186,000 miles per second. It is as if the lines of force that are normally uniformly distributed around a charge become compressed into a plane at right angles to the direction of motion, resulting in a strengthening of the normal electrostatic force, as is suggested in Figure 2 of Chapter 7 by the lines of force around the moving electrons in the double-slit interference experiment.

This relativistic increase in the strength of the force between the two charges, together with a maximum possible velocity for particles equal to that of light, has the crucial effect of leading to a minimum value of the distance to which the two particles can approach each other, regardless of the details of the charge distribution. Even if both charges should be found to have negligibly small size, thousands of times smaller than calculated for a classical electron of finite rest-mass, there would still occur a natural minimum distance between their centers that determined by the values of the charge, the mass and the velocity of light, and not on how much space the charges occupy.

As the relativistic mass of the moving charge seen by the observer at rest with respect to the other charge increases, with its velocity in proportion to the Lorentz contraction factor, thereby increasing the outward centrifugal force, there is an exactly equal increase in the attractive electrical force with increasing velocity, also due to the Lorentz factor. As a result, no matter how big the relativistic mass becomes, the particles remain in equilibrium at a finite distance, equal to the classical size of the electron. It is as if, just as the speed of light limits the speed particles can have relative to each other, the finite size of the electron's rest-mass and fundamental charge, combined with the relativistic contraction that strengthens both the centrifugal and the electrostatic forces, leads to a finite inner size of our physical space. This combination of factors results in a natural lower limit to the distances between the fundamental particles that are sources of electromagnetic fields. If all particles are charged, then our physical space has both outer and inner limits.

In classical Newtonian mechanics as used by Bohr for low velocity electrons in hydrogen, there is neither an upper limit to the relative velocity of particles, nor a velocity-dependent increase in the electrostatic force, so that the expression for the distance of approach of two charges goes to zero as the velocity becomes infinite, producing a mathematically undefined or "singular" state. When the distance between two charges becomes zero, the motional energy proportional to the square of the velocity becomes infinitely large. This is analogous to what occurs when one assumes the electron to be point-like or to have zero size for the source of the electric field, when the energy in the electric field becomes infinite and therefore the electromagnetic mass also becomes infinitely great. The Copenhagen approach to quantum theory could never satisfactorily solve this problem.

If the fundamental particles like the electron have an infinitely large rest-mass, the expression for the minimum approach distance again becomes zero. When one accepts Einstein's assumption that no observer can measure particle velocities greater than the speed of light or transmit signals faster than this limiting speed, it appears that a lower limit to our physical space exists, or that the distances between the ultimate entities can never become zero. This distance is dictated by the finiteness of the mass and charge of the elementary constituents. Since the minimum distance between centers of two charges of magnitude is equal to the "classical diameter" of the electron, taken to be the most fundamental constituent of matter, this is exactly what one would expect geometrically for two "impenetrable" spherical particles just touching each other. Points in a non-Euclidean geometry become small, impenetrable spheres.

A cosmos composed of finite mass particles with finite charges, moving with finite velocities, does not allow for the standard Big Bang model of the creation of the universe, which assumes an infinitely small and dense initial state of things. The possibility arises that Lemaître's "primeval atom" might have had a finite size and a finite density, even it is assumed to contain the entire mass of the universe. If the ultimate particles all have a charge like the electron, the positron and the proton, or if all matter is ultimately electromagnetic, there would not have been a so-called "singularity" at the beginning of the universe, and the present laws of physics can be assumed to have held indefinitely in the past. If the neutron is in fact composed of an electron and a proton, and the mesons that the proton contains are also composed of highly relativistically-moving electrons as suggested by Hofstadter's experiments, then a vast simplification and unification of physical theory becomes possible. There would not be any singularities or infinite quantities.

This amazing result, derived from Einstein's paper on special relativity, provided the crucial answer of how close an electron could come to the proton: the electron would reach a finite distance from the center of the proton, no matter what the actual distribution of charge might be. The geometry of our space appears to be non-Euclidean, both on the scale of nuclear particles and the cosmos. Moreover, since there was now no upper limit to the relativistic mass that an electron could attain in such an orbit, it became possible to believe that the new short-lived states of protons and neutrons are in fact nothing but excited states of structures on a deeper scale, just as Bohm, Vigier, de Broglie and I had come to believe.

My visit to de Broglie's Institute turned out to be more fruitful than I could have imagined. Convinced that there could indeed be a new type of Bohr orbit on the nuclear scale, I asked de Broglie whether he would help me arrange a visit to Bohr in Copenhagen so that I could tell him of this interesting result. At first, de Broglie was hesitant, saying that Bohr would not be happy about talking to someone who had spent so much time in the opposing camp, and who shared Einstein's ideas on the incompleteness of the Copenhagen School's view of quantum theory. But after I explained that surely Bohr would find the application of his own early work to the world of fundamental particles interesting, he relented and wrote a letter to Bohr. When Bohr answered that he would be able to see me, my wife and I made plans to travel via Copenhagen on our scheduled return back to the U.S., leaving the city that we had learned to like so much in November, after fourteen months there.

CHAPTER 9

NIELS BOHR

According to the notes I wrote down after my meeting with Bohr, on Monday November 17th, 1958, the day after we arrived in Copenhagen from Paris, I received a message from Bohr at our hotel that Heisenberg would lecture at the Institute of Theoretical Physics the next morning. Bohr thought I might like to attend that as well as a second lecture in the afternoon. Since I did not want Heisenberg present at my meeting with Bohr, I called Bohr's office and arranged to see him on Wednesday at ten o'clock in the morning.

I walked into the low, rambling building almost exactly on time. At the door to his office, the bushy-eyebrowed, burly face of Bohr smiled a warm, though somewhat cautious welcome.

He motioned me into the spacious room, which had two large armchairs some distance apart at the far end, besides a large desk and a table. After seating me in the armchair away from the window, he settled down in the other one, got out his pipe and stuffed it slowly. Soon he was enveloped in a cloud of bluish smoke, except when on occasions he would slowly get up, walk across the room and sit down again, all the time re-lighting his pipe. During the time we talked, he used up what must have been most of a box of matches in this way, taking a puff and then putting his pipe down again, only to re-light it a few moments later.

Bohr began our conversation by inquiring where I was going. I told him that I was on my way back to the U.S., where I planned to continue my research on the interaction of electrons with solids and the accompanying production of secondary electrons in the outer shells of atoms, an area in which he had carried out some of the earliest theoretical studies. After saying a few words on the elementary character of his own work in this field long ago, he asked me whether he was right in understanding that I had been working on some relativistic extensions of the theory of the hydrogen orbits that Sommerfeld had developed. I realized then that de Broglie must have said something about my findings on the small high velocity orbits of electrons around the proton in his letter to Bohr. But before I could say more than a few words explaining how these orbits were related to the ones Bohr himself had found for the hydrogen atom back in 1913, Bohr indicated that he had asked his colleague Professor Leon Rosenfeld, an expert on relativity theory, to join us, and that he would be coming shortly.

This pattern of our conversation continued for the hour and ten minutes that it lasted. Bohr would ask a brief question, but before I could say more than a few words, he would launch into a long discussion which I soon learned I could not hope to interrupt. Moreover, it was extremely difficult to understand him because a heavy Danish accent made his English sound unlike anything I had ever heard. Since he also often talked with his pipe in his mouth, I missed many of the points in his argument entirely.

Almost at once, the conversation turned into a monologue in which Bohr was visibly agitated and upset. I simply mentioned that I had been trying to obtain a better understanding of the origin of the spin of the electron by pursuing Sommerfeld's 1916 extension of Bohr's theory of the hydrogen atom, so as to consider the elliptical orbits and their precession or change of position in space due to the relativistic effect of motion on the electron's mass. Bohr interrupted me to say that this was quite hopeless and in principle quite impossible. Clearly upset, he got up and walked back and forth across the room while talking with great vehemence, sitting down again after a few minutes to re-light his pipe.

His point was that the spin of the electron was not anything like the rotation of a charge. When I interjected that Uhlenbeck and Goudsmit did try to find a classical analog, he answered that although this was of course a very important and fruitful suggestion, the later development of Dirac's theory using the abstract matrix approach (developed by

Heisenberg) absolutely showed that spin was not just analogous to an ordinary angular momentum, and that it was simply necessary to accept the mathematical rules for handling the quantities in the matrix formalism as evidence that classical physics only held where the quantum of action was so small that it could be neglected.

Bohr then went on to attack the whole approach of Einstein, de Broglie and Bohm as entirely wrong. He became more agitated when he mentioned Bohm, whom he called a fool who dragged out trivial bits of mathematics to cloud the true situation. It was embarrassing for me to see Bohr so emotionally upset about Bohm's work. Bohm, of course, had caused de Broglie to abandon the purely probabilistic approach to quantum theory, something that de Broglie had expressed concern about when I asked him to write Bohr. I tried to interrupt by saying that I too felt that there was a fundamental difference between atomic and classical physics, due to the uncertainty produced by the wave nature of matter and the quantization of charge, mass and angular momentum, but Bohr hardly listened as he insisted that there was no way to visualize phenomena on the atomic scale with ordinary space-time models.

At about this time, Rosenfeld came in. He had hardly sat down when Bohr continued his attack on the efforts of such misguided individuals as de Broglie and Bohm. The vehemence of this otherwise mild-mannered and kindly man really surprised me, and I could sense that Rosenfeld was as embarrassed as I was. It soon became clear that I would not have an opportunity to present my ideas and my effort to reconcile the standpoint of Bohr with that of Einstein and de Broglie. Concerned about Bohr's high degree of agitation, I decided to mainly listen.

Bohr was concerned with something much deeper than physics, which became evident when he handed me a copy of a talk he had just given a few months earlier on the 100th anniversary of the birth of Max Planck, the founder of modern quantum theory. The paper was entitled *Quantum Theory and Philosophy: Causality and Complementarity.* When I read it later back at the hotel, I realized that Bohr's whole philosophy of life was based on an overwhelming need to renounce determinism and detailed space-time descriptions on the atomic and nuclear particle scale in order to achieve a sense of personal freedom, and to accept the mathematical abstraction of quantum mechanics worked out by Heisenberg and Dirac as fulfilling "all demands on rational explanation with regards to consistency and completeness."

I realized that what I was trying to do in particle physics by finding visualizable models for the neutron and the newly discovered mesons was diametrically opposed to what Bohr deeply believed. In the paper Bohr had handed to me he said explicitly that "it seems likely that the introduction of still further abstractions into the formalism will be required to account for the novel features revealed by the exploration of atomic processes of very high energy." He continued by insisting that "the decisive point, however, is that in this connection there is no question of reverting to a mode of description which fulfills to a higher degree the accustomed demands regarding pictorial representation of the relationship between cause and effect."

One of the most interesting remarks made by Bohr while discussing the supposed error of Einstein and de Broglie was that both, but particularly Einstein, always remained detached and lone investigators. In Bohr's Copenhagen group, many minds worked together to perfect the ideas and check them with each other. To me, this was particularly revealing, since the Copenhagen School, based on the philosophy of Bohr and Heisenberg, believed in the fundamental role of cooperative or statistical phenomena involving many particles. In contrast, Einstein and de Broglie emphasized individual entities and the need for a detailed, deterministic, causal description. I was also struck by the fact that Bohr looked down on what he called the "religious belief in harmony" that he felt de Broglie and Einstein maintained. Yet it seemed to me that Bohr had almost the fanatical approach of a fundamentalist preacher, intensely concerned to save my soul from perdition.

Neither de Broglie nor Einstein had felt the need to talk almost continually without making a real attempt to listen, as Bohr did. They were humbler and gentler, much less anxious to defend their views and persuade others of their own ideas with such intensity. I wondered what it was that had made Bohr so adamant in his views, so emotional and on the defensive. Again and again, Bohr made the point that it was impossible to arrive at a more detailed description of phenomena on the scale of atoms and particles than that already provided by the mathematical laws of quantum mechanics. He continuously turned to Rosenfeld for support, often addressing him rather than me.

Many years later, I read in Lewis Feuer's book *Einstein and the Generations of Science* how deeply Bohr had been influenced by the writings of the Danish existentialist philosopher Søren Kierkegaard. I began to realize where this passionately-felt need to renounce a detailed, deterministic description of phenomena originated. Bohr had apparently been introduced to the ideas of Kierkegaard through his professor of philosophy at the University of Copenhagen, Harald Hoffding.

In his lectures, Hoffding had emphasized that "an absolute systemization of our knowledge is not possible," and that "when one thought to extend any analogy based on a part of existence to the whole, no verification is possible." The alternation of analogies was, in Hoffding's view, "the source of the advancement of scientific thought, but no analogy could pretend to provide the model for the totality of Being, since there are always competing type-phenomena." Continuity and discontinuity, according to Hoffding, were standpoints based on such competing phenomena, and neither of the two elements is the only accredited one. The individual personality expressed itself in the choice of method; personality and scientific research were thus co-involved. Hoffding taught that there were hidden regions of the unconscious from which the different philosophical approaches emanated, and that a personal choice was as important in the methodological commitment of scientists and their models as it was among religious philosophers and metaphysicians. The basic problems of philosophy, according to Hoffding, could not be solved, and against this indeterminate background, one made one's choices of complementary alternatives. According to Feuer, these Kierkegaardian ideas taught by Hoffding reached so deeply into Bohr's thinking that he described the atom in its transitions from one stationary state to another as possessing "free choice" like the human subject choosing his qualitative leaps from one stage to another. Thus Bohr's concern was basically with the fundamental problem of ethics concerning the freedom and limitation of will, and he sought to confirm a subatomic world similar in structure with his emotional and philosophical self-definition in the everyday world.

This, then, explained the intense emotional reaction of Bohr to my efforts to develop detailed geometric or "causal" models for the electron, the neutron and the many new particles that were being discovered in cosmic rays and the large accelerators in laboratories all over the world. As Feuer put it, the need to defeat such efforts was linked in Bohr's mind to the affirmation of human freedom. In order to move from the ethical level of existence to the religious, or the highest, an act of renunciation was required. As Feuer concluded, for Bohr the renunciation of strict causality, despite the hardship it entailed, was the essential step to the higher truth of complementarity. One had to renounce the hope of achieving a theoretical system based on any single model, either a particle or a wave. But Bohr's fundamental belief that no description could combine or synthesize the complementary concepts of wave and particle into a single, unified one was diametrically opposite to the task I had begun to undertake.

After a little more than an hour, Bohr fell silent, and I knew he wanted to close the conversation. I said that I had copied the most important equations dealing with the high velocity electron orbits on a few sheets of paper for him to look at, and he replied that he and Rosenfeld would be glad to examine them and write to me. Rosenfeld also asked me to send him the full paper that I was working on, which I agreed to do as soon as it was finished. After a few words about the weather, Bohr shook hands with me and escorted me to the door, ending one of the most memorable encounters of my life.

CHAPTER 10

A VISIT TO FEYNMAN

IT WAS MORE THAN TWO YEARS before I was able to send Bohr a paper on the new highly relativistic orbits that I had arrived at in Paris. When I finally did mail my paper to him, it no longer dealt with the case of an electron moving around a proton, as in the hydrogen orbits that he had studied. Instead of a model for the neutron, I had found a model for the neutral pi-meson, one of the newly discovered particles that was believed to be involved in the force between the protons and the neutrons in the nucleus of atoms. It was a new kind of miniature atom, a hundred thousand times smaller than the ordinary hydrogen atom, composed of an electron and a positron rotating about each other like a double star. It would eventually turn out to be a model for Lemaître's primeval atom.

A conversation with Richard Feynman in June of 1960 completely changed the direction of my work. On the second leg of a trip to try out a new type of image intensifier at the Lowell Observatory in Flagstaff, Arizona, I was to lecture about electronic imaging devices for use in astronomy and particle physics at the California Institute of Technology, to which Feynman had moved from Cornell in 1950.

While I was at Cornell, Richard Feynman had shared an office with my thesis adviser, Philip Morrison. The work on secondary electron emission at Westinghouse that I began in 1952 in an effort to reduce X-ray dose in fluoroscopy led to the development of a new kind of image intensifier and a novel television tube. The tube could accumulate the photons from very faint sources of various types, amplify the photo-electrons they released and store them for long periods like a photographic plate records photons. Thus, it was possible to read out the data electronically, whether from distant stars, from an X-ray image intensifier as used in fluoroscopy, or from small amounts of radioactivity used in nuclear medicine to visualize various organs.

The ability to record single photons, store the electric charges they produce during long exposures and transmit the images electronically like television pictures opened up the possibility of space telescopes that could provide crucial data in support of the Big Bang model. In fact, one of the early orbiting telescopes, operating with this new type of television camera tube sensitive to the ultraviolet photons absorbed by the Earth's atmosphere, provided the first clear spectroscopic evidence for the formation of heavy hydrogen or deuterium in the Big Bang as originally predicted by Gamow. This discovery, by a team of Princeton astronomers led by Lyman Spitzer, also indicated that most of the matter in the universe, if it were to expand to a finite size, had to be "dark" and apparently of an unknown nature, unlike the protons and electrons in the stars.

By the time I returned to Pittsburgh from the Institut Henri Poincaré in November of 1958, much progress had been made by my colleagues Milton Wachtel and Helmut Kanter in developing an image intensifier based on the emission of secondary electrons from a series of thin foils coated on one side with a thin layer of potassium chloride. I had begun to study the properties of this simple salt at the Naval Ordnance Laboratory back in 1947. The image intensifier was designed to sensitize photographic film to very faint objects, where it was crucial to record as many of the photons that entered the aperture of the telescope as possible. Photographic emulsions with their grains of light-sensitive crystals only registered less than one in a thousand of all the photons that reached them, so that it often took many hours to obtain a picture or record a spectrum from a distant star or galaxy. But in the decades since Einstein had developed the theory of the photoelectric effect, research all over the world had led to highly efficient, thin layers of certain metals deposited inside a glass vacuum tube that could intercept photons ten to a hundred times as efficiently as photographic film, each photon leading to the ejection of a photo-electron. The problem was to cause the individual photoelectrons to produce a permanent record in film (or electronically) that would last during the long exposures required to record the photons from distant astronomical objects. Since each of these photons had tens of thousands of times less energy than a single X-ray photon, this was a much harder task than in the case of fluoroscopy.

A number of scientific organizations had begun to sponsor the development of image intensifying techniques that would allow every photo-electron to be recorded by photographic film outside the vacuum tube. One of these was the Carnegie Institution in Washington, D.C. that jointly operated the largest telescope in the world at that time with Cal Tech. The telescope, with a diameter of two hundred inches, was located on Mount Palomar between Pasadena and San Diego. In 1953, the Carnegie Institution appointed a committee to promote the development of electronic imaging methods in astronomy at the prodding of the astronomer William Baum.

When I arrived from Cornell in 1952, efforts then being made at the Westinghouse Research Laboratory to amplify the light produced by X-rays in fluoroscopy suggested to me the idea of thin layers of an insulator such as potassium chloride in a very thin film, no thicker than a few hundred atoms, to multiply the number of electrons entering on one side by collecting the secondary electrons on the other side. The experimental arrangement that I had planned for studying the penetration and back-scattering of electrons could readily be adapted to study secondary emission from the back of very thin foils. By the time I left for Paris in 1957, this approach had shown that collection of four to six secondary electrons leaving the back of a thin layer of potassium chloride was possible for each electron entering the front. With a stack of four such foils mounted on a thin layer of aluminum, and a voltage of a few thousand volts between the foils, a thousand-fold multiplication in the number of electrons could be achieved. By increasing the number to six or eight such foils, the multiplication of the number of electrons would exceed 100,000, and even an ordinary camera could record every photoelectron ejected by a photon from a distant galaxy as a bright flash at the end of the tube striking the phosphor.

The effort to get as many electrons as possible in as few stages as possible led us to discover how to increase the secondary electrons produced in each foil. That in turn allowed the development of a new type of television camera, able to record every photo-electron released and to transmit a signal representing precisely the number of photo-electrons accumulated, from a telescope in high orbit—or a camera on the moon—back to the Earth.

In Pittsburgh, my group was augmented by the arrival of Gerhard Goetze, who had recently come from Germany. I suggested to him that perhaps we could increase the yield of secondary electrons from our potassium chloride layers by preparing them in the form of a highly insulating low density smoke. This would allow the positive charges produced by the emission of secondary electrons to produce an electric field that would pull out more electrons, increasing the amplification attainable per stage. This could be done by evaporating the salt onto the thin aluminum films we were using, not in a high vacuum but in the presence of a small amount of inert argon gas. Goetze set up the equipment, producing a series of snow or "powder-like" forms of thin foil targets that showed five to ten times as many secondary electrons as the solid crystalline films.

We were all very excited about this new development, since it would allow us to greatly reduce the number of stages required to give us the necessary electron multiplication. It improved the fine detail that could be resolved in the astronomical images, since every stage meant a certain scattering of the electrons that would degrade the picture's quality. On the morning after we had been studying a particularly high yield foil in a prototype single stage arrangement, we turned on the voltage and to our amazement the image that we had observed the previous afternoon was still there.

At first, we were discouraged. Such persistence would make it impossible to use a device with foils carrying low density emitters to intensify moving images. But then we realized that if such a foil were placed into the television tubes being developed by another group in our department (headed by Bob Schneeberger) for use in an orbiting telescope being planned by the Smithsonian Astrophysical Laboratory, the photo-electrons produced to form an image could be stored in long exposures and then quickly erased in the process of electronic transmission by scanning with an electron beam.

Schneeberger and Goetze tested this idea within a few days, and found a way to read out the stored secondary electrons rapidly. The results were so promising that Westinghouse received a major research grant from Robert Davis' group at the Harvard-Smithsonian Observatory to build and test a series of the tubes with the low-density targets. They were given the name SEC Vidicons for the secondary electron conduction technique involved. These camera tubes would be used in a number of planned Orbiting Astronomical Observatories able to work in the ultra-violet, launched by the newly formed National Aeronautic and Space Agency (NASA) beginning in 1968. Tubes of the same type later built in England under license to Westinghouse were also flown in the International Ultraviolet Explorer (IUE) satellite launched in 1978, where they operated for more than ten years. By 1988, the SEC Vidicons in the IUE satellite had provided data for 1,571 scientific papers by about the same number of astronomers, a significant fraction of all those active in research in North America and Europe at the time.

In June of 1960, Goetze and I were invited to Arizona to test the performance of our multi-stage image intensifier at the Lowell Observatory in Flagstaff, where Kent Ford and Bill Baum were examining various types of electronic imaging tubes for the Carnegie Institution of Washington. Ford had been working with a tube where electrons were accelerated to pass through a thin window directly into a photographic film. In the years to come, he and my former classmate at Cornell, Vera Rubin, would use electronic techniques to find evidence that all galaxies are surrounded by an enormous halo of dark matter, using the spectral lines displaced from their normal positions due to the rotation of the stars in the disks of spiral galaxies.

Our trials of the transmission-type intensifiers worked quite well, so that this type of tube was subsequently adopted by astronomers at the Kitt Peak National Observatory for their ground-based spectroscopic studies, where lenses and photographic film could be used to record the intensified image on the output phosphor screen.

Baum had arranged for me to lecture on our work in image intensification at the California Institute of Technology following the trials at the Lowell Observatory. I remembered that Feynman had joined the Cal Tech faculty, and so I set up a meeting with him to discuss my work on the nuclear particles. My wife and I rented a car to drive from Flagstaff to Pasadena with our one-year-old son, Dan.

We began the drive from Flagstaff to Pasadena very early in the morning in order to avoid the worst of the summer heat in the Mojave desert, and stayed in a motel that night not too far from Cal Tech, where the next day I was to give a lecture in the prestigious astronomy department about my work in image intensifiers. This was to be one of the most important talks in my professional career, and I knew it. In the morning, I told my wife that she could use the car to do some sightseeing, and that I would take a cab to the lecture hall.

I had called for a cab to pick me up forty-five minutes before the scheduled time for the lecture, but when it failed to arrive after fifteen minutes of waiting, I went to the motel office to ask the owner for his advice. He told me that it was only a short drive, and if the taxi failed to come in another five minutes, he would take me there himself. The five minutes elapsed, and there was still no cab, and so the owner of the motel took me to the campus. Unfortunately, he did not know his way nor did he know in which building the astronomy department was located, and it took me quite a bit of time to get directions. By the time I managed to find the lecture hall at Cal Tech's astronomy department, I was five minutes late. As I approached the door, an older faculty member with an angry face threw it open and stormed past me without stopping. As I entered the room, I realized that the man who had just left the room was the person who was waiting to introduce me, and so I had to introduce myself and apologize for my late arrival, not exactly a great way to start my visit to the Institute. Fortunately, my talk was well received, and I looked forward to my meeting with Feynman to discuss my work on the nuclear particles.

When I called him, he suggested that I come to his office early Saturday afternoon, when there would be no one around, and we could talk without interruption. As usual, Feynman was very informally dressed, and after a few words about our days at Cornell and my reasons for coming to the West, he asked me what I wanted to discuss. I summarized the most recent paper I had prepared, attempting to explain some of the new particles. Using the method for calculating force originally suggested by Einstein in 1905, I argued that the electrons in high-velocity or relativistic Bohr orbits could explain the excited states of proton. An electron and a proton would reach a minimum separation on the order of the electron's classical size, regardless of the charge distribution in either particle, as I had discovered in Paris.

Feynman agreed that this was interesting, and so I described how, for these highly relativistic velocities close to that of light, the orbits would not remain stationary in space, but would precess or rotate as a whole, just like the orbit of Mercury around the sun, as illustrated in Figure 10.1 below.

I pointed out that this precession, already discovered by Arnold Sommerfeld in 1916 for the hydrogen atom, would take up half the total angular momentum. As a result, if one were to quantize the total angular momentum, half of it would be associated with the motion of the electron in its orbit, and half with the rotation of the frame in which the orbit is at rest relative to the observer.

Again, Feynman had no problem with this result, and I continued with my idea that if such states existed in the nucleus, this electron would have a large enough mass and therefore a short enough de Broglie wavelength, due to its enormously high velocity, to be identified with the meson involved in the nuclear force as originally suggested by Yukawa. I then described that if one assumed the proton and the heavy electron formed a kind of molecule with a fixed distance between them due to a

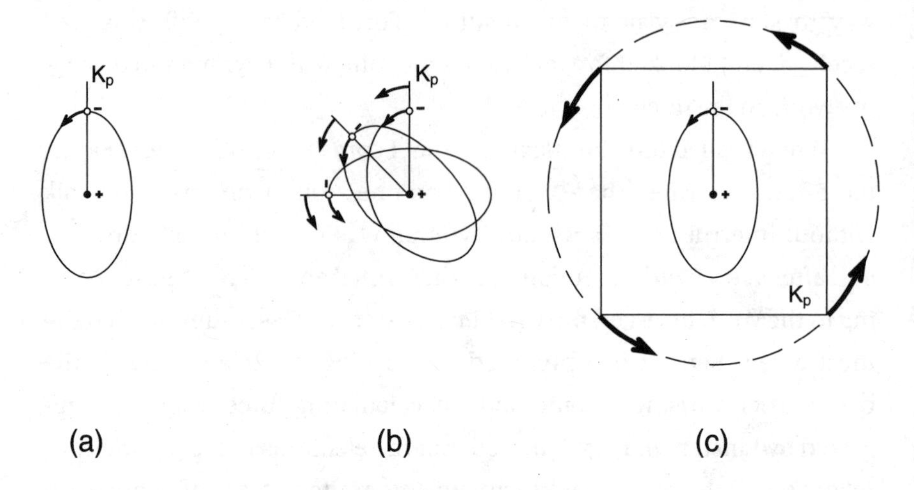

Figure 10.1 Schematic diagram illustrating precession of an orbit. (a) A classical orbit of an electron around a proton stationary in space (b) An orbit rotating or precessing as a whole as a result of relativistic motion (c) The same orbit as in (b) as seen by an observer at rest in the rotating reference frame marked K_p relative to which the precessing orbit is stationary.

minimum approach distance, the usual excited rotational states of such a system seemed to fit the masses of some of the very short-lived new particles referred to as "resonances," and similar states might be formed by an electron-positron system.

But this time Feynman's response was very different. He completely upset me by saying that these proton-electron molecular states that I had mentioned were "a lot of bull," and asked me whether I had worked out the simpler case of the electron and positron. When I answered that I had not yet done it, he said impatiently: "Goddamn it, go to the board and work it out."

I spent much of the rest of the afternoon working out the relativistic orbits of the electron-positron system on the blackboard of Feynman's office under his watchful eyes, subjected to occasional comments. But when it turned out that the relatively crude calculation gave a mass about 274 times that of the electron, Feynman said: "You so-and-so, you have just found a classical model for the neutral pi-meson."

Instead of being delighted, I left Feynman's office absolutely crushed. I had labored for more than ten years, whenever I could find the time away from the work on my "cobbler's job" of secondary emission and image intensification, trying to find a model for the neutron, and now Feynman had thrown it all out as worthless.

That weekend, in the motel room in Pasadena, our son took his first steps. Under Feynman's prodding, I too had taken the first steps towards an explanation of the evolution of matter before the creation of neutrons and protons at the moment of the Big Bang. I had found that there was no upper limit to the energy or mass of the allowed electron-positron states, because the attractive force kept increasing with the repulsive centrifugal force, as the relativistic mass increased—that is, as the velocity approached the speed of light, due to the contraction of the electron's source and the resulting increase in strength of the electric field in the direction at right angle to its motion. Such a purely electromagnetic electron-positron pair would never fly apart and therefore could be energized to a state high enough to contain the entire mass of the universe.

Nonetheless, at the time I was deeply discouraged by what had happened to my hopes for explaining the new heavy particles. I drove to Stanford where I told Hofstadter, with whom I shared many common interests, of my experience with Feynman. I knew that I would have great difficulty in getting a paper on a classical electromagnetic model for a meson published, given the dominance of the abstract Copenhagen approach to theoretical physics among the reviewers. Most scientists felt simple classical models were useless on the nuclear particle scale. I proposed to ignore forty years of quantum field theory. It seemed totally hopeless. But Hofstadter replied that if I worked out all the details of the model Feynman had forced me to examine, he would review it, if necessary with Morrison who had by then moved to MIT, and submit it for me to the *Physical Review*.

When I returned to Pittsburgh, for the next few months I spent the mornings working at home on the electron pair model and pursuing the experimental work on electron physics at the laboratory in the afternoons. To my amazement, I found that I could not only obtain agreement with the most recent mass value to better than a half percent, but I was also able to calculate the observed lifetime of the neutral pi-meson within the experimental uncertainty. When I was finished, I sent the paper to Hofstadter. I could not believe my eyes when a few months later I received a postcard from the editors of the *Physical Review* that it had been accepted for publication without my being asked to make any changes.

By the time the paper appeared in the July 1, 1961 issue, I had begun to work on the extension of the electron pair model for the neutral pi-meson to the heavier mesons then being discovered as composed of two, three or more pairs of these relativistic states forming short-lived systems analogous to ordinary atomic scale molecules. These systems, in fact, had the same type of rotational excited states that I had originally attributed to the electron-proton states I had shown to Feynman. But now, there was a clearly established model for such excited rotational states that not only explained their masses but also their sizes, spins and modes of decay with only the need for the four basic atomic constants: the charge of the electron e, its mass m_o, the speed of light c, and Planck's Constant h divided by 2π—the angular momentum of the electron pair system. I realized that with my wrong model involving an electron and a proton I had found the right form of the molecular-like states that fitted the pattern of the newly discovered heavier mesons.

Since the scattering of high energy electrons against protons in Hofstadter's laboratory had just revealed a structure consistent with the masses and sizes of systems composed of a few neutral pions, it suddenly seemed possible that one could understand the proton itself as nothing but a particularly strongly bound, and thus long-lived, structure composed of relativistically moving electron-positron pairs held together by a central positive charge or a positron of highly relativistic mass.

If this were indeed to be the case as a result of further studies at higher energies, one could hope that the neutron would turn out to have the same basic structure as the proton to which an electron of high relativistic mass could attach itself briefly. Lorentz's idea of the purely electromagnetic nature of all matter would become a reality. There could then be an ideal fluid ether in which photons would be a quantized version of Helmholtz's classical vortex rings with internal motions at the speed of light, and all of matter would be a form of light since photons of sufficient energy could turn into electron-positron pairs. The rest-mass of electrons and positrons would be nothing but a manifestation of their localized, internal spin or rotational energy, giving a simple interpretation to Einstein's relation that energy equals mass times the square of the speed of light, $E = mc^2$.

Realizing that these highly relativistic electron-positron systems were the analogs to short-lived, hydrogen atom-like low energy "positronium" states (consisting of an electron and positron orbiting around a common center but some 100,000 times larger). I arranged to see the scientist who first studied these phenomena: J. A. Wheeler, Feynman's teacher at Princeton and co-author with Bohr of the theory of uranium fission.

In a paper published in 1946, Wheeler had considered the possibility that the mesons might be composed of a large number of such positronium states, each with a mass close to that of two electrons. I thought he would be interested in the analogous high mass relativistic states for mesons that I had found on the scale of nuclear particles. Indeed, during our conversation, Wheeler did in fact express an interest in the new high mass states, and suggested that I use another method of calculating the interaction between the rapidly moving electron and positron that he and Feynman had been studying shortly after World War II, and see whether it would give the same results.

The method of calculating the forces Wheeler referred to involved what are called "half-retarded and half-advanced potentials," a somewhat unorthodox and more complex technique of calculating the interactions between moving charges. The "potentials" have their origin in the work of German scientists in the middle of the nineteenth century that competed with Maxwell's approach. I had used a third approach, that of Lorentz and Einstein. Because it implied abandoning causality, or the usual rule that a cause must precede its effect, I had not studied this first method, and so I discussed the problem Wheeler had posed with Alfred Schild, a mathematician interested in relativity theory and my colleague at the Westinghouse Research Laboratory. At my urging and that of Wheeler, who knew him well, Schild undertook this calculation a few years later at the University of Texas. Contrary to what I had found, Schild concluded that with this technique of calculation, the relativistic electron-positron system did not result in a mass greater than two electron masses, counter to the conclusion Feynman had expressed to me after I had finished the derivation on the blackboard in his office, and counter to what Hofstadter and other reviewers of my paper had concluded.

When I examined the paper Schild published in the *Physical Review* in 1963, I realized why he had come to this conclusion. He had in effect carried out his calculation in a reference frame that was at rest in the laboratory, in which the large motional energy of the particles is exactly canceled by the large electromagnetic binding energy, whereas in actuality, this frame rotates rapidly due to the precession as discussed above. When one considers that the energy associated with the motion of precession equals that of the electron relative to the rotating frame, one does indeed arrive at the high positive energy of 274 electron masses, explaining the mass of the pi-mesons as Feynman had agreed, and allowing one to predict correctly the sizes, masses, and spins of the heavier mesons as I had done in a paper published at a conference on nucleon structure at Stanford the same year that Schild had published his paper.

When I had an opportunity to visit Wheeler in Texas some twenty years later, he could not accept the idea that Schild, who had tragically died at a relatively young age, might have made a mistake. Wheeler never changed his mind, even though by this time the relativistic electron pair model allowed me to explain both the masses of the mesons and the many other new particles and the masses and sizes of the cosmological systems. I could only speculate on the reasons for his attitude.

Perhaps it was because over the years he had been deeply committed to the views of Bohr and Heisenberg, having spent a few years with Bohr at his Institute both before and after World War II, and collaborating with him on such crucial papers as the one that predicted that only uranium-235 but not uranium-238 would fission readily under bombardment by slow neutrons and thus lead to a self-sustaining chain reaction.

But I was not unprepared for Wheeler's intransigence. In *The Structure of Scientific Revolutions,* Thomas Kuhn discusses the difficulty with which scientists of a given generation can accept a new approach or paradigm. He quotes a particularly perceptive passage by Darwin at the end of his *Origin of Species:* "Although I am fully convinced of the truth of the views given in this volume . . . I by no means expect to convince experienced naturalists whose minds are stocked with a multitude of facts all viewed, during a long course of years, from a point of view directly opposite to mine. . . . [But] I look with confidence to the future—to young and rising naturalists, who will be able to view both sides of the question with impartiality." Kuhn also added the example of Max Planck, the founder of quantum theory, which in its own early period had so many opponents. Planck sadly remarked that "a new scientific truth does not triumph by convincing its opponents and making them see the light, but rather because its opponents eventually die, and a new generation grows up that is familiar with it."

Wheeler's approach to the conceptual problems of physical theory were similar to those of Bohr, with whom he had formed a close friendship. He shared Bohr's view that the role of the observer in quantum physics is crucial in bringing an object into existence—a point the fallacy of which Einstein illustrated to me when he had asked me whether the tree in his garden was still there when I turned my head away. So I was also not surprised that after I had sent Bohr a copy of the paper dealing with the relativistic electron pair model for the neutral pi-meson, Bohr sent me a letter in which he raised a point of criticism that seemed to him to make such a model in principle impossible. Bohr pointed out that the radius of the orbit in my model was only a quarter the size of the classical radius of the electron and positron, much too small to allow the two particles, assumed to have their finite, classical theoretical "shell-model" size, to describe such a small orbit about each other.

Curiously, as in the case of Schild's paper, this objection also turned out to be connected with the neglect of the rotation of the reference frame in which the orbit is at rest relative to an observer. As Einstein had explained long ago in an address to the Prussian Academy of Sciences in 1921, six years after he had formulated his theory of general relativity, in a rotating frame of reference, ordinary Euclidean geometry no longer holds. Space becomes curved, giving rise to gravity: "In a system of reference rotating relatively to an inertial system, the laws of disposition of rigid bodies do not correspond to the rules of Euclidean geometry on account of the Lorentz contraction."

I had recognized this phenomenon while I was working on the electron-proton orbits in Paris. The precession of the highly relativistic states would not only cause the electron to shrink in the tangential direction along its direction of motion around the proton, but the enormously high rotational velocity and the resulting inward acceleration would lead to a locally very high distortion or curvature of space, analogous to the distortion of the normal uniform density of air by a tornado. The pressure of the air would be reduced near the center of such a vortex, an analogy that holds particularly closely if all matter is indeed electromagnetic, since this allows a liquid-like ether to exist as Lorentz had shown. Such a high local space distortion produces a locally high gravitational force, as would be observed if one were to send a photon or a small spherical particle through this distorted region, causing it to be deviated inward as if there were a force acting on it, just like a golf ball circles a cup on the green of a golf-course. Indeed, during a total solar eclipse in 1919, the observation of the deflection of photons passing near the edge of the Sun as a result of the distortion of space by the Sun's large mass, exactly as predicted by Einstein's General Theory of Relativity convinced the scientific world of the validity of Einstein's theory on the origin of gravity.

This local distortion of the "flat" Euclidean space was shown by Einstein to lead to a shrinking of any object or measuring rod in the radial direction at right angle to the orbital motion, and a flattening of the electric field that it produced, causing the force it exerts on another charge to increase as discussed above. I had calculated how much a spherical object would shrink in the radial direction in the case of the Earth orbiting in the gravitational field of the Sun. I found that it shrinks by exactly the same amount in that direction as it flattens, due to the Lorentz contraction in the direction of motion. This meant that in any gravitational or inverse square-law field, an initially spherical object is restored to its fully symmetrical, spherical shape in an equilibrium orbit by virtue of the non-Euclidean or distorted form of the local space, although with reduced radius. Since all objects, including measuring rods, are also reduced in size, and clocks are slowed down by the same factor, an observer in the rotating reference frame can not detect the rotation of the reference frame in which he is at rest by any physical measurements or experiments.

This was the essence of Einstein's General Theory of Relativity. It extended the fundamental idea that an observer should always find the basic laws of physics to be independent of his or her state of motion, not only in the case of a reference frame in uniform linear motion but also in a uniformly accelerated frame of reference such as a rotating frame, where the direction of the velocity rather than its magnitude is constantly changing. The same situation exists in linearly accelerated reference frames, as when a car is speeding up at a constant rate. The result was just the same as in the case of linear motion, where the Special Theory Relativity led to a contraction of lengths and a slowing down of clocks that makes it impossible for the observer in the moving system to detect any uniform, non-accelerated motion relative to any other uniformly moving frame since he would simply interpret the effect of a uniform acceleration as a gravitational field. Likewise, such uniform motion relative to an absolute reference frame such as the ether of Descartes, Newton and Maxwell could not be detected in Michelson and Morley's experiments with light beams on the Earth moving through the ether. This inability to detect an ether led the young Einstein to derive the relativistic laws of motion without assuming the existence of an ether or any absolute rest-frame.

When I calculated the amount of radial contraction of an initially spherical electron around a positive charge, I found that since there exists a minimum approach distance as there does for the Earth around the Sun in general relativity, the amount by which the electron shrinks in the radial direction is exactly equal to the shrinkage due to its tangential motion. Thus, a large local distortion of space, such as that produced by a rapid rotation of the frame in which an observer rotating with it is located, appears to act like a locally high gravitational field, causing the electron to be restored to a spherical shape, as if it were at rest, only reduced in radius by the Lorentz contraction factor.

In the particular case of the lowest allowed state of the relativistic electron-positron pair system, the reduction in the diameter of the electron's classical charge distribution was by about 274 times, equal to the relativistic increase in its mass. Remarkably, in his 1921 address to the Prussian Academy, Einstein discussed the case of a high local space curvature as produced by the rotation of a reference frame on the scale of atomic particles:

> For even when it is a question of describing electrical elementary particles constituting matter, the attempt may still be made to ascribe physical importance to those ideas of fields which have been physically defined for describing the geometrical behavior of bodies which are large as compared with the molecule. Success alone can decide as to the justification of such an attempt, which postulates physical reality for the fundamental principles of Riemann's geometry outside the domain of their [original] physical definition.

Indeed, in a paper presented at an earlier meeting of the Academy of in 1919, Einstein had considered the possibility that a high local degree of space-curvature or a high local gravitational field resulting from an abnormally high value of the gravitational constant within the electron might account for its stability. The idea that a high local gravitational field could exist inside the electron and within the immediate neighborhood of an electron-positron system would turn out to play a key role in the early phase of the evolution of matter before the Big Bang, or before the creation of the protons and neutrons together with electrons of ordinary matter as we know it.

The success of the relativistic electron-positron model in accounting for the large mass of the pi-meson, its size, its lifetime and the strength of its interaction with other mesons sufficient to explain the great strength of the nuclear force by about the same factor, 274, clearly argues for Einstein's view that one must not give up the efforts to find detailed space-time or causal descriptions for phenomena on any scale, no matter how small. Moreover, this same electron-positron model would lead me to an explanation of the masses of all the other heavier mesons that had been discovered by the early 1960s as rotationally excited states of two or more such pair-systems forming molecular-like structures, determined only by the value of the fine-structure constant α and no other arbitrary or adjustable parameters. Moreover, as I was to learn in the years to come, this number turned out to be crucial not only in determining the masses of all the other nuclear particles and the strength of their interactions in terms of the mass and charge of the electron, but it also proved to be the key to the mass and structure of the universe.

No stable particles of finite rest-mass other than electrons, positrons and protons have ever been observed to be the end-products of collisions in high energy particle experiments. That argues in favor of detailed Bohr atom-like models involving only electrons and positrons forming molecular-like systems on a scale 100,000 times smaller than ordinary atoms and molecules. Thousands of experiments with accelerators have revealed no other stable particles were formed in the decay of the many new particles. That test has continued to hold up, despite a thousandfold increase in the available collision energy provided by modern particle accelerators since they were first used in the 1930s.

Bohr died in November 1962. I was saddened by the death of one of the greatest scientists of our age, one who had devoted himself for decades to warning the world of the danger of nuclear weapons that he had helped to develop in the hope that their existence would end war forever. He had also worked for cooperative international control of the peaceful uses of nuclear energy, arguing that without complete openness of all research laboratories in this field, the danger of a nuclear arms race that could lead to the destruction of civilization would be an ever-growing threat. A few days before his death, his warning had come perilously close to reality when the Cuban Missile Crisis nearly started a nuclear war.

I never did discuss with him why two fundamental charges could describe an orbit smaller than their own classical diameter. His idea of quantized orbits, when joined with Einstein's ideas on the origin of gravity, turned out to be the key for understanding the structure of mesons and all the other nuclear particles. Bohr always had had enormous admiration for Einstein, despite their diametrically opposed personalities and philosophies, and he would probably have been willing to consider the idea that in a rapidly rotating reference frame, where a non-Euclidean geometry had to exist according to the theory of general relativity, the electrons would have been reduced in size as seen by a stationary observer in the laboratory, making such small orbits possible after all.

Fortunately, other physicists were extremely supportive of my efforts. De Broglie, who at age seventy had just retired from the directorship of the Institute Henri Poincaré, sent me a letter in which he thanked me for the papers I had sent him, saying that he was pleased to see me continue with my intriguing research. "You have not let yourself be stopped by the obstacles that you have encountered," he wrote.

De Broglie continued in his position as permanent secretary of the French Academy of Sciences and lived a long and active life well into his nineties. In June of 1964, a few years after his retirement, it gave me particular pleasure to send him another paper that Hofstadter had agreed to submit to the international physics journal Nuovo Cimento, of which Hofstadter had become an associate editor. In it, I showed how the charged mesons as well as the heavier mesons fitted the electron pair model, and that extending de Broglie's suggestions on neutrinos would support a fundamental unity of physical phenomena in electromagnetic terms. De Broglie thanked me very graciously for sending him this paper and said he intended to discuss it with his colleagues.

CHAPTER 11

LABORING UNDER THE NUCLEAR SHADOW

SUPPORT CAME FROM ANOTHER SOURCE, namely from a physicist, Seth Neddermeyer, who had participated in the discovery of both the positron and the first of the mesons in cosmic ray experiments three decades earlier. Apparently, Neddermeyer, a professor at the University of Washington in Seattle, had long believed that the mesons were composed of the electrons and positrons into which his cloud chamber studies had shown them to decay, and he congratulated me on finding a model for the neutral pion.

When Oppenheimer assembled thirty or so scientists at Los Alamos in April 1943, he included Neddermeyer, his old student. It was Neddermeyer who first proposed and pursued in detail the idea an atomic bomb detonated by imploding a spherical shell of explosives. That explosion would then compress a hollow shell of plutonium into a dense ball of metal able to produce a "chain reaction." A few initial neutrons would cause uranium or plutonium nuclei to fission, releasing 200 million electron volts of energy each as they did so. (In ordinary beta decay, an electron typically is ejected with the energy it receives by being accelerated to one million volts or less.) This unprecedented amount of energy released from a single nucleus was sufficient to make a whole grain of sand jump a few millimeters into the air. But the fission of a heavy nucleus also produced about two more neutrons. These in turn resulted in another generation of nuclei disintegrating and releasing yet more neutrons, until by multiplication this process produced explosive power equivalent to thousands of tons of TNT in a few millionths of a second, using only a few pounds of fissionable material.

Despite negative reactions from Oppenheimer and Bethe, Neddermeyer persisted in promoting this method as the most likely to allow speedy construction of an atomic bomb in time to end the war. It turned out that he was right. It was this type of implosion bomb that was successfully detonated the morning of July 15, 1945 near Alamogordo, New Mexico, releasing an amount of energy estimated to equal ten thousand bombs, each producing the equivalent of about 2,000 pounds of high explosive in a blinding flash, or a total of twenty million pounds of TNT.

Over the next few years, I often met with Neddermeyer at scientific meetings, discussing not only the nature of nuclear particles, but also how to educate politicians and the public about the need to end nuclear testing and the nuclear arms race before a nuclear war resulted. Neddermeyer, like most of the scientists who had worked at Los Alamos, and especially those who had been personally involved with the design and construction of the bombs, felt a deep guilt about his part in the devastation of Hiroshima and Nagasaki.

A number of the physicists who worked in the Manhattan Project, including Philip Morrison, who had had the task of assembling the plutonium core of the implosion bombs, had joined with Einstein to form the Federation of Atomic Scientists. Their goal was to educate the public and keep fission under civilian control. I had joined the FAS in my days as a graduate student at Cornell, where Morrison had become one of its most prominent spokesmen. As a result of his many talks and articles warning of the danger of nuclear war, he had been viciously attacked by Senator Joseph McCarthy, for whom anyone who argued for nuclear disarmament was a Communist and a traitor to his country.

By 1962, the danger of nuclear war had greatly intensified, and government officials were urging the construction of fallout shelters as a way to survive a nuclear war. A number of us in the Pittsburgh chapter of the FAS decided to study recently declassified information of the effects of nuclear weapons to find out whether shelters were really useful. Since I was involved in work to reduce the biological risk from relatively low doses of diagnostic X-rays in fluoroscopy, I volunteered to examine the danger from nuclear bomb fallout. As I have described in detail in my 1972 book, *Low Level Radiation,* as well as in *Secret Fallout* written ten years later, through my research I found a disturbing study reprinted in Congressional hearings. This paper by Dr. Alice Stewart, an epidemiologist at Oxford University, led me to conclude that even the small amounts of radiation associated with worldwide fallout from the testing of nuclear weapons were likely to have already produced a significant increase in childhood leukemia all over the world. Dr. Stewart had discovered that children exposed in their mother's womb to two or three diagnostic X-rays just before birth had double the normal risk of developing leukemia before age ten. Calculations by members of the FAS found that a single fifty megaton bomb detonated by the Soviet Union in the fall of 1961 produced the equivalent of an abdominal X-ray to every person in the northern hemisphere. Thus, there was no doubt in my mind that there would be a very significant increase in childhood leukemia throughout the world in the next few years.

At that time, the FAS was urging a test-ban treaty, and so I decided that the information on the unexpectedly great effect of small doses of radiation, magnifying by ten to one hundred times the risk of childhood leukemia or other forms of cancer, could help to end nuclear testing if it were to become widely known.

By the time of the Cuban Missile Crisis in October of 1962, I had finished the draft of a paper on the implications of continued nuclear testing for the newborn. With the help of a number of concerned scientists such as Barry Commoner, I managed to get my paper accepted for publication in the journal *Science,* despite its initial rejection by the editor, Philip Abelson. I had also sent an advance copy to the White House. Kennedy's science adviser, Jerome Wiesner, apparently briefed the President on its content since Kennedy was planning to announce his intention to negotiate a treaty to end above-ground testing, and he needed public support to get the treaty ratified in the Senate. I learned this some months later, from the librarian of the British Foreign Office who requested a copy of the paper on which Kennedy had been briefed.

Early in July, the new director of the Westinghouse Research Laboratory, William Shoupp, called me into his office and asked whether I would like to take an all-expenses paid vacation with my family in Europe in August. I was wary of this unusual offer from the man who had designed the reactor core for the first nuclear submarine, and I asked what was up. He replied that he had heard from the Westinghouse vice president in charge of nuclear reactor sales that there would be Congressional hearings in August on the effects of low-level radiation where I would be asked about my paper just published in *Science*. Shoupp suggested that perhaps I could simply submit my testimony in writing. After talking it over with my wife, I told Shoupp that since I would be representing the FAS and not Westinghouse, if such an invitation were in fact to come, I would have to testify in person in order to be able to defend my findings. He thereupon said that the laboratory would pay the expenses for my and my wife's trip to Washington.

At the hearings of the Joint Committee on Atomic Energy, the well-known Harvard epidemiologist Brian MacMahon, who had also been invited to testify, was present. He had just published an independent study in the prestigious *Journal of the National Cancer Institute* that supported Stewart's findings. MacMahon agreed with my conclusion that there was no evidence for a safe threshold below which no damage would occur. He also concurred that the developing infant was much more susceptible than the healthy adult to low levels of radiation, and that there was reason to believe that even the relatively small doses from nuclear testing could produce an increased risk of childhood leukemia and cancer. Moreover, he agreed that there was no evidence for any safe threshold of radiation, certainly at least down to the relatively small dose from a single abdominal X-ray. As I knew from my study, this dose was hundreds of times smaller than government plans allowed people to receive after leaving a shelter following a nuclear attack.

A few days earlier, Edward Teller, the Hungarian theoretical physicist who had come to the U.S. to join George Gamow at George Washington University, and who had been the driving force behind the development of the hydrogen bomb, had testified before the Senate Foreign Affairs Committee against ratification of the test-ban treaty. He had argued that further testing was essential to the development of an anti-ballistic missile system that could protect the nation from nuclear missiles, and that there was no danger from the present levels of fallout. Thus I was greatly relieved when towards the end of September the Senate ratified the treaty to end all atmospheric testing that Kennedy and Krushchev had signed. I would finally be able to return to my work on the lunar scientific station, the electronic techniques of reducing radiation doses in radiology, and the problem of understanding the structure of the proton and neutron, as well as the relation between the electromagnetic and nuclear forces that now looked so promising.

Towards the end of June 1963, shortly after my article in *Science* appeared, Neddermeyer and I had met at a large conference on nucleon structure at Stanford where the Nobel Prizewinner Hofstadter had been doing pioneering work in the field. Neddermeyer gave a paper on recent cosmic ray experiments in which positive mu-mesons with a mass of about 207 electron masses, a net angular momentum or spin of $1/2\ \hbar$ and seven billion electron volts' worth of energy showed an anomalous behavior in collisions with electrons in carbon targets. He interpreted the results as possibly indicating that in the course of these collisions a neutral particle was formed similar to the neutral pion, but with a net angular momentum of one quantized unit $\hbar$, instead of zero.

Such a neutral spin 1 particle, twelve electron masses heavier than the spin 0 state, was required to exist by my theory due to a repulsive magnetic force between the electron and positron in the spin 1 case, when the magnetic moments, acting like small bar-magnets, are parallel as shown in Figure 11.1, but it had not been definitely seen so far.

Since this spin 1 particle was neutral, it would not leave a track in a cloud chamber. In addition, it was predicted to decay relatively slowly either into two neutrinos or three gamma rays, so that it would escape from the cloud chamber and be difficult to detect. No convincing evidence had been found for such a spin 1 neutral pion, where the spins of

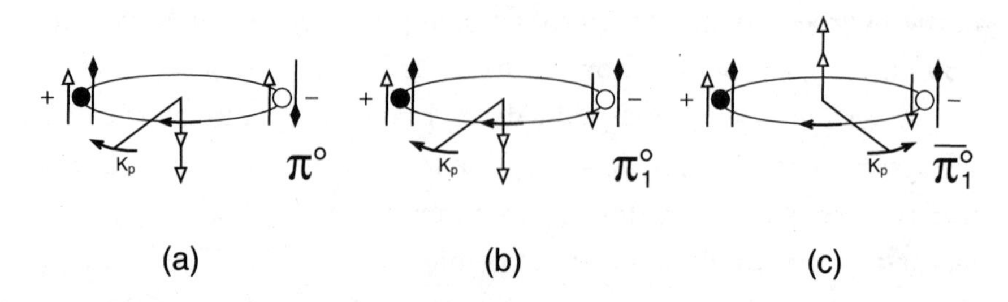

Figure 11.1 Schematic drawing of the three possible distinct types of relativistic electron-positron systems with spins shown as open arrows and magnetic moments as solid arrows. Each type consists of a negatively charged electron and a positively charged positron, moving around their common center of mass at rest in the laboratory. The two particles each have a spin angular momentum of (1/2) h/2π, which can be oriented either parallel or anti-parallel. Their orbital angular momentum is always equal to twice the spin, or h/2π, and it can be either in the same sense as that of the electron, or opposed to it.

(a) The net angular spin 0 neutral pion observed in the laboratory by its decay into two gamma rays.

(b) One of two possible types of net spin angular momentum 1 neutral mesons not observed in their lowest energy state, because they mainly decay into two neutrinos. The orbital motion is in the same sense as the spin of the electron, as shown by the direction of the open arrows, each of which represents a unit of (h/2π).

(c) The other spin 1 system, but with orbital motion in the same sense as the spin of the positron, identical in mass but like a mirror reflection of the system shown in (b). The magnetic moment associated with each of the spinning charges acts like a small bar magnet, with the black arrow the north-pole. When the magnetic moments are parallel, they repel each other, decreasing the strength of the binding energy and thus increasing the net mass of the spin 1 pions, in accordance with Einstein's famous relation $E = mc^2$, from 264 to 276 m_o electron masses, where m_o is the electron rest-mass.

the positron and electron point in the same direction but are opposed to the angular momentum due to their orbital motion, and so I was encouraged by Neddermeyer's results.

This much more stable spin 1 meson in its ground-state was apparently the basic building block for all the newly discovered heavier molecular meson states, and therefore also the major component of the proton. I had presented a preliminary paper on the molecular models for the various mesons at a Physical Society meeting in New York in January of 1963, and I hoped to give a paper at the Stanford meeting in June since this would be published in the proceedings. I knew how difficult it was to get past the reviewers in the establishment journals in this area. I even had some difficulties in getting the chairman of the theoretical session at the Stanford meeting to accept my unorthodox paper, but fortunately Hofstadter intervened and I was able to present my findings.

As the paper in the proceedings showed, the simple model used only the pi-meson mass derived from the fundamental atomic constants and the relativistically increased electromagnetic binding energy—which explained the strong nuclear force. It fitted the nineteen molecular-like system masses observed up to this time to within less than five percent without any adjustable parameters. The observed new mesons fitted amazingly well to exactly the same kind of rotational states that applies to ordinary chemical molecules consisting of two, three, four and five pions arranged in regular geometric structures with a constant distance between nearest neighbors. But such simple, classical space-time geometric models were totally out of fashion in the dominant Copenhagen approach, as indicated in the proceeding's opening address delivered by the highly regarded theoretical physicist Geoffrey Chew. Referring to the problem of the "structure" of nuclear particles, Chew said:

> The term 'structure' in fact is misleading in that it may connote some kind of detailed spatial distribution of matter inside the nucleon. The combination of principles of relativity and quantum mechanics implies that no meaning can be attached to such spatial

> distribution. What are observed in all experiments discussed in this volume (as in all other work on nuclear physics) are scattering amplitudes as functions of the momenta of ingoing and outgoing particles. The collection of all scattering amplitudes is called the S-matrix, and to know the S-matrix is to know all that can possibly be known about the sub-atomic world.

The deep schism between Einstein and the Copenhagen School was starkly put by Chew when he ended this paragraph as follows:

> The existence of a sub-atomic space-time continuum need not, and perhaps should not, be assumed.

Here was the essence of Heisenberg's abstract mathematical approach that Bohr had accepted, diametrically opposed to the views of Einstein, de Broglie, Bohm, and Vigier that I had used. Yet two pages later, Chew admitted that "the characteristic symmetries associated with strong interactions may have a dynamical origin, to be revealed at the same moment that one understands the properties of the observed particles." But for the theorists of the Copenhagen School, there was a deep-seated psychological resistance to visualizing dynamic space-time models that might reveal the origin of the symmetries and the scattering phenomena observed in the experiments. The discovery of such "hidden variables" as Bohm was convinced existed would prove that quantum mechanics as formulated by Heisenberg and Bohr was after all not "complete" or the end of the road in our understanding of the physical world. Bohm's view was that of Einstein, resulting from his conversations with him while he was in Princeton. It was strongly supported by the philosopher of science Karl Popper. As Popper put it some years later, the Copenhagen doctrine regarded the task of science as to provide useful tools for making calculations. Popper contended that what science seeks is "truth, approximations to truth; explanatory power, and the power of solving problems; and thus understanding."

Most theorists at the time felt that there were no truly fundamental building blocks of matter, and that every strongly interacting particle could be regarded as composed of every other one in a so-called "particle-democracy." As a result, Chew argued that in the effort to understand the properties of the proton, one "cannot expect ever to achieve theoretical calculations of a precision comparable to that of the calculations of atomic physics." Thus, there was a sense of a growing crisis among theoretical physicists. There seemed to be no end to new "elementary" particles being discovered, a despair and a loss of hope that a true understanding of phenomena on the nuclear particle scale could ever be achieved. But I now had reason to believe that there were indeed fundamental building blocks of matter, and that they were nothing other than the familiar electron and its positively charged anti-particle the positron, all closely related to each other, the direct descendants from the heaviest particle of all, Lemaître's primeval atom.

I shared Popper's view of the task of science, and remembered Einstein's advice to be "stubborn." And so, with the encouragement of de Broglie and Neddermeyer, the support of Hofstadter and the manager of my department, John Coltman, together with that of the director of the Westinghouse Research Laboratory, Bill Shoupp, I set out to find detailed models for the remaining nuclear particles.

CHAPTER 12

STRANGE SCIENCE

TO UNDERSTAND HOW A SINGLE PRIMORDIAL ATOM was at the origin of matter and the universe, I had to understand the detailed structure not just of the neutral pi-meson and its molecular-like states, but of all other particles, including the charged mesons and the proton. It was also necessary to understand the way these systems explained the strong and weak nuclear forces for the idea that matter is composed of nothing but electrons and positrons to be credible and that all forces could be understood in electromagnetic terms.

I set out to explain the mu-meson that Neddermeyer and Anderson had discovered. The mu-mesons or muons, as they were widely called, represented a particularly tricky puzzle. They seemed to interact with matter more like a heavy electron or positron of 207 electron masses. However, instead of being stable like the electron, they decayed to electrons or positrons, depending on their charge, together with two neutrinos, all in about two millionths of a second. The muons seemed to play no role at all in the strong nuclear force as the somewhat heavier pi-mesons or pions did, and there was no explanation for them in current particle theory. The muons were found to have exactly the same spin $\hbar/2$ as the electron, and not a spin of zero as did the charged pions, or $\hbar$ like the neutral pions. They were always found to either carry a positive or a negative charge, unlike the pions that also occurred in neutral form. Moreover, the muons were only produced whenever a charged pion with its larger mass of about 273 electron masses decayed, accompanied by the emission of a neutrino.

In terms of the Bohr-type model that had worked so well for the neutral pion, it seemed clear to me that the positively or negatively charged muon had the same structure as the charged pion. With a charged pion, an electron or a positron is bound to a neutral pion, but with different amounts of energy and angular momentum. The charged pion, being heavier (containing more energy), would be the particle in which the added charge is in a greater rotational energy state than in the muon. The pion decays to the less heavy muon, emitting a quantum of radiation that carries away the difference in energy and angular momentum in the form of a neutrino. Similarly, in the hydrogen atom the difference in energy and angular momentum between a greater and smaller energy state of the orbiting electron is carried by a photon. The difference in the strength of the interaction of a pion and a muon with the mesons in the protons and neutrons would then have to be due to their different mass, de Broglie wavelength and spin, since the emission of a neutrino of spin 1/2 changes the pion from a spin 0 or 1 unit of $\hbar$ to a lighter particle of half-integral spin $\hbar/2$.

I discussed this idea with Neddermeyer, and it would agree with the latest experimental findings that he had reported at the Stanford Conference. I found that an electron or positron could indeed be bound to the spin 1 form of the electron pair system shown in Figure 11.1 (b) or (c) of Chapter 11 by virtue of the fact that this system had a non-zero magnetic moment. That made it behave like a magnet which attracts another particle such as an electron or positron, as shown schematically in Figure 12.1 below.[1]

When I finished my calculations, I was stunned. By being able to calculate both the mass and lifetime of the mu-meson, using ordinary electromagnetic theory, I had apparently stumbled upon a confirmation of de Broglie's conjecture, namely that the neutrino was a form of electromagnetic radiation or a kind of photon—but one that carried an angular momentum only half that of photons. If I was correct, I had found a connection between the so-called "weak" nuclear force and the electromagnetic force. In 1967, this link was arrived at by a totally different route by two highly regarded theoretical physicists, Steven Weinberg and Abdul Salam, who used abstract symmetry theory. Once again, as in the case of

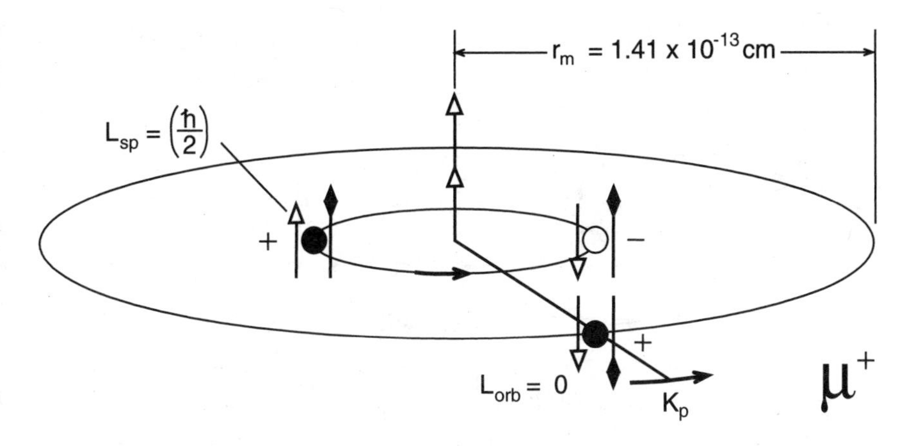

Figure 12.1 Schematic model of the positively charged mu-meson. The core consists of a spin 1 neutral pi-meson, composed of an electron and positron orbiting each other with angular momentum $\hbar$ whose sense of rotation is parallel to that of the positron. The positron in the outer orbit is at rest relative to the precessing frame of the central pair system (indicated by the line K_p) in which orbits are closed and the laws of classical mechanics apply. Open arrows indicate the sense of rotation according to the right-hand rule where the thumb points in the direction of the arrow and the fingers indicate the sense of rotation. The solid arrows represent the direction of the magnetic field associated with the spin of the charges, like a small bar magnet with the north pole where the solid arrowhead is.

atomic structure, it appeared that very different approaches led to similar conclusions.

Then I turned to the charged pi-meson, treating it as a muon in a higher state of angular momentum and energy, just as if it were a Bohr-type structure, but with an angular momentum higher by half of the unit $\hbar$ assumed by Bohr, since it decays to the lower mass muon with the emission of a neutrino of spin $\hbar/2$ relative to an observer in the precessing frame of the central spin $\hbar$ neutral pi-meson, as shown in the diagram of Figure 12.2 below.[2]

Not only did these semi-classical dynamic models give surprisingly close numerical values between the calculated and observed masses and lifetimes, but they explained the existence of two different types of neutrinos. One type was produced when muons decayed, and another type was produced along with the emission of an electron, as occurred

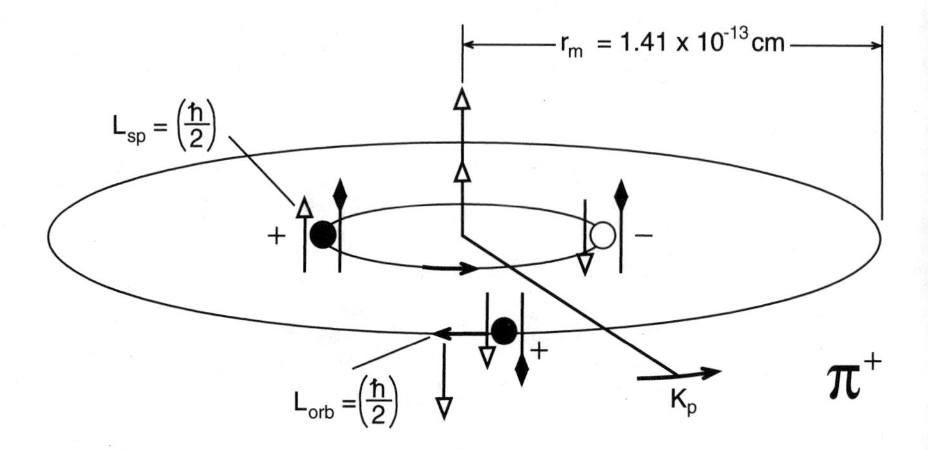

Figure 12.2 Schematic diagram of the positively charged pi-meson model. The system is identical with the mu-meson model in Figure 12.1 above, except for the added orbital angular momentum of the positron in the outer orbit relative to the precessing reference frame of the central spin 1 neutral pion core. Open arrows indicate the sense of the angular momentum by the right-hand rule; the closed arrows indicate the direction of the magnetic field north pole where the solid arrowhead is.

in the decay of a muon. This discovery had been made in 1962 in high energy accelerator experiments by a Columbia University group at the Brookhaven National Laboratory. It was found that when high energy muons decayed, the neutrinos emitted only produced muons when striking protons, but never electrons. In terms of the models for the muon and charged pion, the two neutrinos emitted together with an electron in the decay of a muon each carry away a half unit of spin, as a result of the central spin 1 pair annihilating into two radiation quanta (to conserve linear momentum). But when the pi-meson decays from its higher angular momentum state to the lower angular momentum state of the mu-meson, it does so with the emission of just one neutrino that has an angular momentum of spin 1/2 relative to the precessing frame, but almost no angular momentum as seen by a laboratory observer. For the observer at rest in the laboratory, it carries only the absolute minimum value 137 times smaller than $\hbar$ /2. The number 137 is a close approximation to the inverse of the fine-structure constant.[3] It is this fundamental pure or dimensionless number that determines the mass of the neutral pi-mesons that I had worked out in Feynman's office, close to twice 137, and it enters as the determining factor into the mass of all the other unstable nuclear particles and resonances. Moreover, it determines the strength of the relativistic force between the electron and positron in the pi-meson state, and thus the strength of the strong nuclear force that is associated with the pi-meson. And since the proton appeared to contain two K-mesons each composed of two pions, I knew that this same fundamental number of quantum theory would enter into the theoretical value for the mass of the proton and neutron.

I also calculated the lifetime of the neutron and found a value of 900 seconds, of the same order as the best observed value of 1,132 seconds when I did the calculation in 1964, which later came down to 898 ± 16 seconds, a surprisingly close degree of agreement, contrary to what Chew had deemed possible for nuclear particles.

I showed these results to Hofstadter at the April 1964 meeting of the Physical Society in Washington, and told him that I would have no chance of getting a paper with my semi-classical models past the theoreticians used by the *Physical Review* as referees. He agreed to go over the paper himself and submit it for me to the European journal equivalent to the *Physical Review*—*Nuovo Cimento*, of which he had become an associate editor.

A few weeks later I received a letter from Hofstadter telling me that he had submitted the article for publication, and the only concern he had was that it was rather long. But within a month or so a postcard arrived from Italy, and to my great relief the article was accepted for publication in the January 1965 issue without my having to shorten it.

I received numerous requests for reprints of the article, and it was favorably reviewed in the *European Nuclear News*. I also looked forward to the chance to present my latest findings at a major conference scheduled for June 1965 at Ohio University in Athens, Ohio, on the extension of the electron pair theory to the structure of the proton and the many new excited states that kept fitting the theoretical energy levels that my

Nuovo Cimento paper had predicted. Evidently, many experimentalists were attracted to the idea that the chaos in particle physics might be replaced by the simple idea that the electron and positron alone may be the truly stable, indivisible building blocks of all matter. But, as Neddermeyer had told me, when he spoke to one of the most highly respected theoreticians in the world whom he knew well from his days at Los Alamos about the basic idea of my paper, he refused to even look at it.

In the invitation for the Ohio University Conference on Resonant Particles, the chairman, B. A. Munir, had asked for copies of the papers to be submitted in advance so that the proceedings could be printed and made available as soon as possible in this rapidly developing field. This served as a spur to put my latest results on record, particularly on the extension of my *Nuovo Cimento* paper to the problem of the structure of the baryons, a term that included the proton, the neutron and the newly discovered unstable heavy particles that decayed to protons or neutrons called "hyperons" and "hyperon resonances." This would allow me to circumvent the battle with unsympathetic reviewers in this field.

The fact that the molecular model for the excited meson states, or "meson resonances" as they were called, fitted all the states that decayed into mesons discovered since I had sent the *Nuovo Cimento* paper to Hofstadter, and that similarly spaced resonances were found for the excited states of protons and neutrons as "nucleon resonances" were important clues to their structure. These excited states were found when pi-mesons were fired against protons or neutrons, and there were peaks in the plot of the number of pi-mesons scattered versus the energy of the pions. Such peaks suggested that at certain energies, the meson had formed a temporary compound with the proton or neutron that lasted for thousands of revolutions, rather than being a simple collision in which particles bounce off each other, as shown in Figure 12.3 below.

When I compared the energy of such "compound states" as shown in Figure 12.3 (a) and (b)with the rotational states of my molecular models consisting of pions, I found that those for three pions fitted those for a certain group of eight nucleon states to within less than three percent.

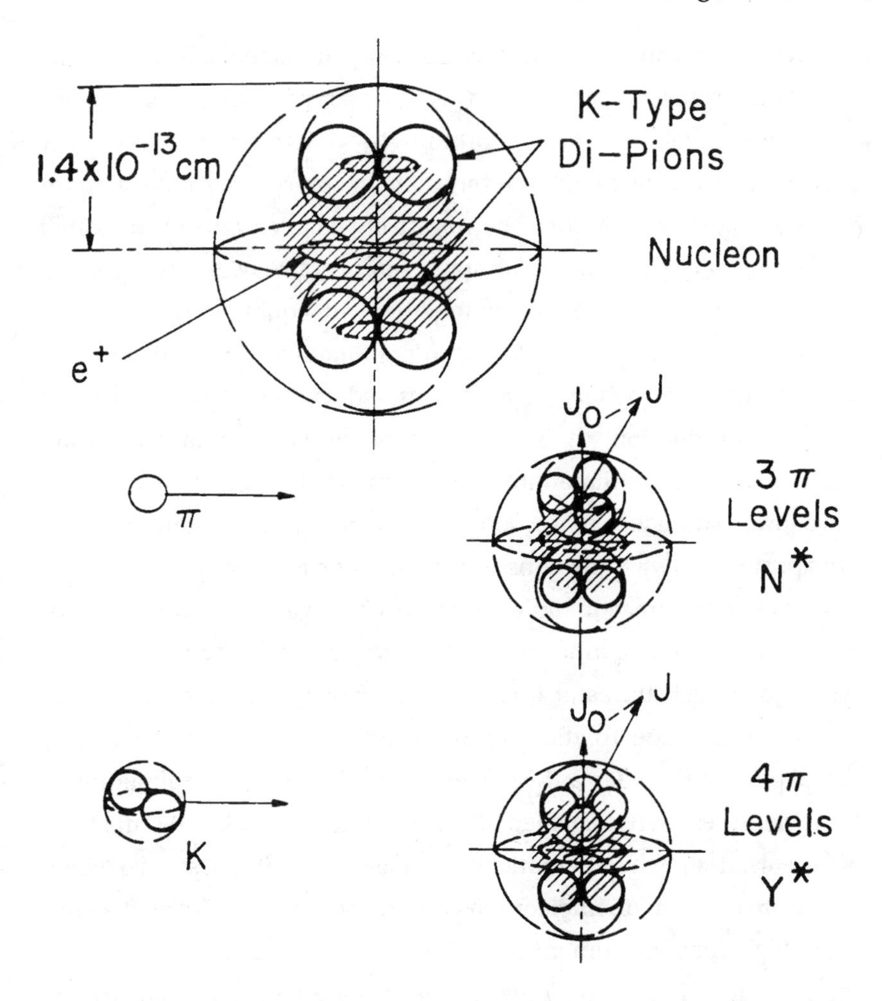

Figure 12.3 Schematic diagram of proton, (a) before temporary capture of a pi-meson, (b) after capturing a pi-meson, and (c) after capturing a K-meson composed of two pions to form so-called hyperon resonances. When a collision leads to four pions at each side of the proton rotating in opposite directions, hyperons are produced which have the same spin as the proton in (a) because the net angular momentum of this structure is the same as that of the proton due to opposite senses of rotation of the set of four pions at each end, canceling their contribution to the total spin or angular momentum of the symmetrically excited structure.

Moreover, for all but two of the theoretically predicted states up to the highest energy investigated, there was an observed resonant state with a pi-meson emerging from the proton. By that time, it was widely accepted that protons contained mesons in some very stable state for the following simple reason. The negatively charged anti-protons that had first been created in the laboratory in 1955 by Owen Chamberlain and his research team at Berkeley, together with equal numbers of positively charged protons, were found to annihilate into mesons and not into pure radiation when they encountered ordinary protons a short time after production, unlike the case of the electron and positron, which normally annihilate into two gamma rays. Also, the size of protons allowed some four of my neutral spin 1 mesons to exist in the space of the proton, or two K- mesons each containing two pions.

As the table of meson molecular states in the article in the proceedings of the conference indicated, both a two-meson and a three-meson system had closely the same total mass in their lowest state of zero angular momentum due to the stronger binding force for three closely packed pions, referred to as K-mesons. Thus, the simplest explanation of the nucleon resonances, called N* at the time in which only one pion was involved shown schematically in Figure 12.3 (b), was that when pions were incident on nucleons at certain energies, they formed a temporary high angular momentum three-pion state with two of the pions that normally existed in their lowest energy state at one end of a proton .

In rare cases, a collision can lead to one K-meson consisting of two pions being temporarily added to one side of the proton as in Figure 12.3(c), leading to a different group of states with different energies or masses and different spins called Y^*. Since all the various unstable particles involving the proton are seen to be based on the same basic electron pair systems of pion mass, the fact that the calculated masses fitted the observed values within a few percent argued strongly that all stable and unstable nuclear particles have a close family relationship to the first such pair in the universe, the Lemaître atom.

The fundamental unity of all forms of matter and forces therefore seemed to be understandable in the kind of detailed space-time dynamic models that theorists like Chew regarded as inherently impossible. But the very similarity in basic structure of ordinary molecules makes it clear why the same group-theoretical concepts developed earlier for molecules seemed useful in grouping and classifying the new high-energy particles, and made mathematicians like Chew confident that one did not need anything more than group theory and the scattering matrices of Heisenberg.

It was impossible for group theory to provide the mass of the various particles such as the simplest pi-meson or the K-meson consisting of just two neutral pions. Yet with the use of detailed space-time models such as had been so successfully used in atomic and molecular physics, it was possible to calculate the masses of these particles within a few percent of the observed values. Thus, the K-meson consisting of two pions had a theoretical value of 974 $m_{0,}$ or electron masses, somewhat more than half the mass of a proton of 1836 m_0. Thus, it seemed to support the hypothesis that the proton was composed of two K-mesons, stabilized by a strongly bound relativistic positron exchanged between them. In the same general manner, two protons are bound together by a low velocity electron to form a positively charged hydrogen molecule or ion on a scale 100,000 times larger, as shown in Figure 12.3(a) above. Moreover, a rough calculation of the binding force produced by a relativistic electron of mass comparable with the mass of two pions gave a mass for the proton within three percent of the observed value.

This model of the proton was further supported by the observed spacing of the different hyperon states, in which K-mesons sometimes formed strangely long-lived structures when they interacted with protons at certain energies, for which they were given the name "strange particles." By 1964, the four such states that had been discovered and which ended up as protons and mesons fitted the spacing of the rotational states of four neutral pions very closely, forming another type of K-meson of nearly the same energy as the two and three pion systems. This close family or "group" relationship among these different particles was further evidence that they were constituted of basic entities arranged in different geometrical arrangements, just as ordinary atoms and molecules are composed of the same basic entities, electrons, protons, and neutrons.

But this was not all the evidence for an electron-positron model of the nucleons and mesons. There was yet another set of excited nucleon states called "hyperons" or "strange baryons," described in Figure 12.3, because they had strangely long lifetimes compared with what was expected, about a trillion times longer than the time it takes for a particle to make one orbit around proton. Moreover, despite the fact that they seemed to be complex particles created in collisions of protons with K-mesons, they had the same minimal angular momentum of $1/2\,\hbar$ of a proton. As mentioned in the discussion of Figure 12.3, I found that this puzzle could be solved if one assumed that in the case of the hyperons, two sets of K-mesons of the type consisting of two pions became temporarily attached to the proton structure, one at each end and rotating in opposite directions. In this way, their angular momenta would cancel, leaving the net angular momentum at the value $\hbar\,/2$ of the proton. When I compared the theoretical mass values for the lowest excited states of four-pion type K-mesons bound to the mass of the proton, the resultant masses for all four known hyperon states, the lambda, sigma, chi and omega hyperons, fitted the observed masses to within a few percentage points.

The amazing thing about these results was that just as for the original model of the neutral pi-meson, there were no adjustable or arbitrary parameters needed to fit the masses; the only quantities needed were the four fundamental constants e (energy), m_0 (mass), c (speed of light), and $\hbar$ (Planck's Constant divided by 2π).

The hyperons had first been recognized as part of a family of "strange particles" by the young theoretical physicist Murray Gell-Mann in about 1953, who assigned a "strangeness number" to them whose sum was preserved in the course of high energy collision processes. By 1960, he and others had found that the baryons could be organized into a group of eight arranged with six in the form of a six-sided hexagon and two in the center, on the basis of abstract group theory, a pattern which he had dubbed "the eight-fold way." Moreover, he had found a similar grouping for the mesons, so there was group-theoretical evidence based on symmetry for some sort of dynamic relationship among the mesons and baryons before 1960 when I worked out the neutral pion mass in Feynman's office, which I later learned happened to be next door to Gell-Mann's.

By 1961, Gell-Mann and independently another theoretician, the Israeli physicist Yuval Ne'eman, had found that the symmetry group denoted by SU(3) appeared to govern the pattern of the strongly interacting nuclear particles. The following year, Gell-Mann predicted that the symmetry of the eight-fold way required the existence of a fourth hyperon which he called the omega minus, whose charge, approximate mass and other important properties he was able to specify. Shortly thereafter the omega-minus was found, with the same properties as Gell-Mann predicted. For this and other contributions to physics, Gell-Mann was awarded the Nobel Prize.

At the Conference on Resonant Particles in June of 1965, Ne'eman's paper entitled "Band spectra generated by non-compact algebra" preceded my presentation. As the title indicated, since the term band-spectra had first been used for the light emitted by rotating molecules, the group-theoretical models now being applied to nuclear particles had originally been developed for ordinary molecules, very similar to the ones I had arrived at based on the relativistic electron-positron model for the neutral pion. The paper was based on recent work he and Gell-Mann had done to understand the newly discovered meson and baryon excited states. They had used the operator algebra approach discovered by Heisenberg for calculating the intensity of light emission for the various energy levels in the hydrogen atom, and extended by other theorists to the rotational excited states of the strongly interacting particles. Ne'eman arrived at a general formula for these states very similar to mine, but without a specific dynamic model for the constituents, he could not present any definite mass values for the various types of particles or explain their physical relationships, a problem that is inherent in the abstract group theoretical approach.

After some ten minutes of discussion in which Ne'eman admitted that there is no basic reason why any particular group such as SU(3) should be applicable, it was my turn. After briefly summarizing the underlying model for the mesons in terms of relativistic electrons and positrons, I showed a graph of the molecular rotator levels with their constant spacing, creating the various levels for systems of 2, 3, 4 and 5 pions, and how well the latest particles fitted the predicted levels for each type. Then I discussed how the hyperon levels came about physically, and how simply they explained that their spin was the same as that of the proton.

There followed a discussion during which I was asked to explain in more detail how these new kinds of positronium states arose, and why they had not come out of earlier theories, such as those of Dirac. Then Ne'eman asked me about the decay of the muon and pion, and whether the neutrinos are bound in the mesons. I explained that they were created in a radiation process similar to the case of photons, only that they had a smaller spin and were radiated slowly by much heavier and smaller electrons than in the atomic case. Then Ne'eman asked me a question Feynman had asked him: "Would you be ready to drop your theory if the 720 Mev meson turned out to have a spin 0 and not 2?"

"A theory of this kind either has to fit every particle, or it is not good at all," I answered. Ne'eman replied that that was "certainly a satisfying answer."

I realized later that my answer to Feynman's question was really a little rash, remembering how Einstein had said that a single experimental observation that does not agree with his theory of relativity would not cause him to abandon it, especially since quite frequently initial experimental observations may turn out to have been wrong, as they actually were in the case of the variation of the electron's mass with its velocity.

My model for the proton and the hyperon—for which Gell-Mann had introduced the SU(3) symmetry based on three quarks—was actually consistent with the idea of three distinct types of entities making up the proton structure and the structure of the mesons at which I had arrived. One was the K-type mesons containing two neutral spin 1 pions making up the outer two parts of the proton. The second type of heavy component was the massive positron in the center of the proton and neutron providing the strong bond between the two structures, while the third was the massive electron that Heisenberg believed was exchanged between the proton and the neutron, converting one into the other. But just how the positron managed to provide such a strong force to hold the parts of the proton together despite all attempts to disintegrate it, insuring its stability through billions of years since the Big Bang despite the fact that it contained charges that normally tended to annihilate each other in a fraction of a second, was to remain an unsolved problem for years to come.

Primarily to tackle the problem of the detailed structure and stability of the proton I decided to spend a semester at Stanford with Hofstadter at the end of 1966. And I had less incentive to remain at Westinghouse: the drain of the Vietnam War on government resources caused cutbacks in funding. Among the hardest hit were the follow-up missions in NASA's Apollo lunar exploration program. That put an end to the promising lunar scientific station program on which I had been working.

What I regretted most was the loss of the chance to have a small twelve-inch remotely-controlled telescope equipped with our SEC Vidicon operating on the moon. With its long exposures during fourteen day-long lunar nights free from any atmospheric disturbances and the ability to resolve very fine detail, ten times finer than could typically be achieved on Earth due to the turbulence of the atmosphere, it had the potential of outperforming the largest Earth-based telescopes in its ability to resolve details of the very distant galaxies evolving near the edge of the universe. Working with two astronomers long interested in instrumentation for remotely operated instruments, Martin Harwit and Stephen Maran, and the engineers at the Goertz Optical Company in Pittsburgh, we had developed the detailed design of a telescope that weighed only 160 pounds on Earth so that with the lower gravity on the moon, it could be readily set up by two astronauts. It was primarily intended to explore the suitability of the lunar surface as a base for future astronomical observations with larger telescopes. However, with the newly developed SEC Vidicon capable of accumulating electronic images for many hours with a dynamic range of 10,000 to 1 for point sources that had already been tested and accepted for the orbiting observatory program and the hand-held camera for the lunar landing, there was a chance to see fine detail that would not come again until the large telescope proposed by Lyman Spitzer at Princeton University—with whom we had been working—could be placed into orbit.

I felt that the ability to establish the form of galaxies in the early universe was so important because it would not only decide the choice between the steady-state and the Big Bang theories, but it could also decide between two different versions of the Big Bang which differed on the origin of galaxies and stars. The predominant view was that the initial explosive formation of matter led to an expanding hot cloud of gas from which galaxies formed gradually over billions of years by the collapse of regions that happen to be slightly denser than others. This theory was based on the work of the British astrophysicist James Jeans, developed in the late 1920s to explain the formation of galaxies and stars. But a small group of astronomers, mainly associated with the Armenian astronomer Victor Ambartsumian, believed the scenario developed by Lemaître in the early 1930s. According to Lemaître, the universe did not start as a hot gas, but as an extremely dense atom-like seed that he called the primeval atom, which divided to produce the seeds of smaller systems such as galaxies, stellar clusters and stars, and only in the last step would all the smaller seeds lead to the formation of ordinary matter.

In this view of the Big Bang, galaxies would be expected to begin as very compact dark seeds of very high density that first produced small spherical galaxies. Later, they ejected spiral arms which eventually formed large thin disks. Also, smaller objects such as dwarf galaxies should have their origin in the center of large galaxies, and Ambartsumian had found many examples where smaller galaxies seemed to have been ejected from larger ones in chain-like arrangements. As a result, being able to see the form of very distant and thus newly forming galaxies and the pattern of their arrangement could decide whether Lemaître's theory of the origin of galaxies from massive seeds or the widely-accepted theory of origin by gravitational collapse was the correct version of the Big Bang. Moreover, with my conclusion that all matter was ultimately formed of relativistically rotating electron-positron pairs, it seemed to me increasingly likely that the primeval atom of Lemaître might turn out to be an enormously massive electron pair system.

Thus, it was a very deep personal disappointment to face the cutback of funds after the first few Apollo flights. It would not only end my hope for an early lunar surface telescope capable of resolving the shape of distant galaxies, but it would also delay the large orbiting telescope that was expected to use our camera tube, a telescope that had a real chance to decide the question of the age and origin of the universe.

At that time, my group had become involved in experiments using the SEC camera tube to record very low dose images produced by gamma rays in nuclear medicine with Dr. Wang-Yen at the University of Pittsburgh. These turned out to be rather promising, and after I had presented our recent findings at a conference at the School of Medicine, the head of the Department of Radiology, Dr. Elliot Lasser, asked me whether I would consider leaving the Westinghouse Research Laboratory to direct a new laboratory for the development of new imaging techniques. The laboratory was then under construction with a grant from the National Institute of Health and would be finished in 1967. Since I was still heavily involved in the lunar scientific station program, I said that I would need to see what was going to happen with the drop in NASA funding, and he agreed to wait a few months.

To leave Westinghouse and direct full-time a group working on electronic imaging in medicine would be a major decision. After discussing it with my wife, I decided to ask the director of the Research Laboratory, William Shoupp—to whom I was reporting on my work on the lunar scientific station—whether he would allow me to accept an invitation to spend the fall semester with Hofstadter at Stanford working on the structure of the proton. Shoupp very graciously agreed to give me another sabbatical leave. With our two young children, we flew to San Francisco in August of 1966, near to where Stanford University was located.

By the time I arrived at Stanford, Hofstadter had begun to study the low-lying excited states of the proton with his linear accelerator, and it occurred to me that there might be a state where the two sets of pairs of electrons in the model of the proton as shown in Figure 12.3(a) would rotate as a rigid body. The energy involved was relatively small, and so there was a chance that it might be found experimentally. Hofstadter agreed to look for it, but a careful set of experiments failed to find it, and I was rather discouraged. Years later I realized that the stability of the proton that made our universe possible required that the two halves of the proton rotate independently about an axis joining, them as in the hyperon, so that the individual components could absorb energy of any amount. At low impact energies, it would be the molecular-type rotations of the two pions that composed the K-meson structures on each side that absorbed the energy, while at greater energies, this would happen by internal excitation of the individual electron pair systems, which could take up any amount of energy, no matter how large. This absorbed energy of excitation would then lead to the formation of pairs of electron-positron systems that would be emitted as mesons, while the proton returned to its normal state without disintegrating. But the details of this whole process and the particular structure that allowed the proton to maintain such incredible stability was not to become clear until many years later, when I began to understand the process by which the Lemaître atom divided and ultimately gave rise to stable protons at the time of the Big Bang.

By January 1967, when it was time to leave Stanford, I had decided to accept the position at the University of Pittsburgh. The application of electronic techniques to X-ray imaging and nuclear medicine seemed to me to have an enormous future, both in reducing radiation doses and in applying modern computer technology to the enhancement of images for earlier and better diagnosis. I had become convinced that eventually film could be replaced by electronic images stored on magnetic tape and disks, as well as in computer memories. These techniques could revolutionize radiology and allow new and less invasive surgery under the guidance of stored X-ray images. Lasser's offer was an exciting opportunity to apply the technology that had been developed in the space-program to imaging in medicine.

There was also another consideration in making this decision, related to my involvement in the struggle to end nuclear testing and the continuing build-up of nuclear weapons. At that time, a further escalation in the arms race was threatening to take place, namely the construction of an anti-ballistic missile system that would use nuclear-tipped missiles in an attempt to destroy incoming warheads. Remembering the findings of Dr. Alice Stewart, I realized that the detonation of the numerous atomic explosions in the atmosphere that would be involved in such an effort would produce enormous amounts of fallout, endangering the future of our society by its effect on the newborn, even in the unlikely event that the system worked perfectly and no enemy missiles reached their targets. But to bring this to light would be much more difficult if I were to stay at Westinghouse than if I were at the university, where I would have tenure as a full professor in the School of Medicine and could speak out far more freely. Thus, the decision to leave the Research Laboratory seemed the best course of action, although I had enjoyed working there for over fifteen years.

The combination of organizing a new laboratory at the university and becoming increasingly involved in the battle to end the threat of nuclear war, as described in *Low Level Radiation* and *Secret Fallout*, meant I had to give up my work in nuclear particles and cosmology for many years. A few years later, I learned that large amounts of radioactive fission products were routinely released from commercial nuclear reactors, and I publicly warned about the dangers to human health. I became estranged from many of my colleagues in the field of nuclear physics, who had hoped that the peaceful atom would atone for Hiroshima and Nagasaki and, perhaps, implicitly for their own sins of having brought on the shortlived "Nuclear Age."

Seven years after leaving Stanford, I was at last able to return to work in fundamental particles and cosmology, when I learned about the discovery of a new kind of meson while attending a meeting in Washington in November 1974 sponsored by Ralph Nader's Public Citizen anti-nuclear organization, Critical Mass. It was a particle whose properties convinced me it was a massive excited state of the electron-positron model that Feynman had made me work out, and that its highest state could indeed be Lemaître's primeval atom from which the particles of matter and the structure of the universe had evolved.

CHAPTER 13

NAMING THE FIRST PARTICLE OF THE COSMOS

WHAT WAS IT ABOUT THE NEW MESON that made me suddenly believe that Lemaître's primeval atom was a single electron-positron pair? Which properties of this particle made it possible to think that at long last it might become possible to understand how the universe began and how ordinary matter originated?

To begin with, the J/Psi particle, as it would soon be called, was the most massive single nuclear particle found so far, and thus the densest object known to exist. It was approximately the size of an ordinary nuclear particle like the other mesons, as could be deduced from its absorption in matter. At the same time, at 6,060 electron masses, it had more than three times the mass of the proton, which was in turn 1,836 times heavier than the electron, so that the new meson had a greater mass per unit volume than any other form of matter. But this was precisely what Lemaître's hypothesis required: a new state of matter much more dense than any known in his time, one that would allow not only the mass of a star or a galaxy to be compressed into the space of a proton, but the mass of the entire universe, estimated to contain billions of galaxies. And because there was no upper limit to the energy contained in the relativistic electron-positron pair system that Feynman had me work out on his blackboard fifteen years ago, I realized that a higher energy version of this microscopic structure could in principle form the seed of stars, galaxies and the entire universe, as difficult as this was to contemplate.

In fact, it was the absence of either theoretical or experimental evidence for such a dense state of matter that had prevented most astronomers from agreeing with the highly respected Armenian astronomer Victor Ambartsumian when, at a scientific congress held in Brussels in 1958, he suggested that small galaxies appear to be ejected from the center of larger galaxies, and that the spiral arms of galaxies form by the ejection of matter from a massive central seed along the lines suggested by Lemaître a quarter of a century earlier.

Despite its enormous mass, the J/Psi existed thousands of times longer than would be expected for such an energetic particle. Clearly, if the trend of lifetime increasing along with increasing mass were to continue, then by the time such relativistic electron pairs had the vastly greater masses of the seeds of stars and galaxies, enormously long lives would be associated with them. Long after the Big Bang, the lives of such massive pairs would have to last long enough to form the seeds of cosmological structures such as dwarf galaxies, recently ejected from the center of ordinary galaxies.

What made it likely that the J/Psi was indeed a high energy state of the relativistically rotating electron-positron system, and thus the same type of entity as the Lemaître atom, was the fact that the precisely defined measured mass of 6,060 electron masses was extremely close to the theoretical mass of the twenty-third level of the spin 0 form of the neutral pi-meson. And since the new meson was seen at the Stanford Linear Accelerator in a colliding beam of electrons and positrons, and not in an accelerator where a proton hits a solid target, as was the case with the discovery of the same particle at Brookhaven, there could be no doubt that the J/Psi and all the various particles into which it decayed arose from the interaction between two pure electromagnetic charges. It seemed that the J/Psi was completely analogous to the model of the Lemaître atom from which all matter in the cosmos originated.

The fact that the J/Psi was formed at the energy where the two possible types of neutral pi-mesons had exactly the same allowed, discrete or quantized energy levels could hardly be coincidence. And so I looked for a possible molecular-like rotational level of two, four, or eight neutral pions that would have nearly the same energy into which the initial electron-positron pair could decay by a series of divisions by two, and which would

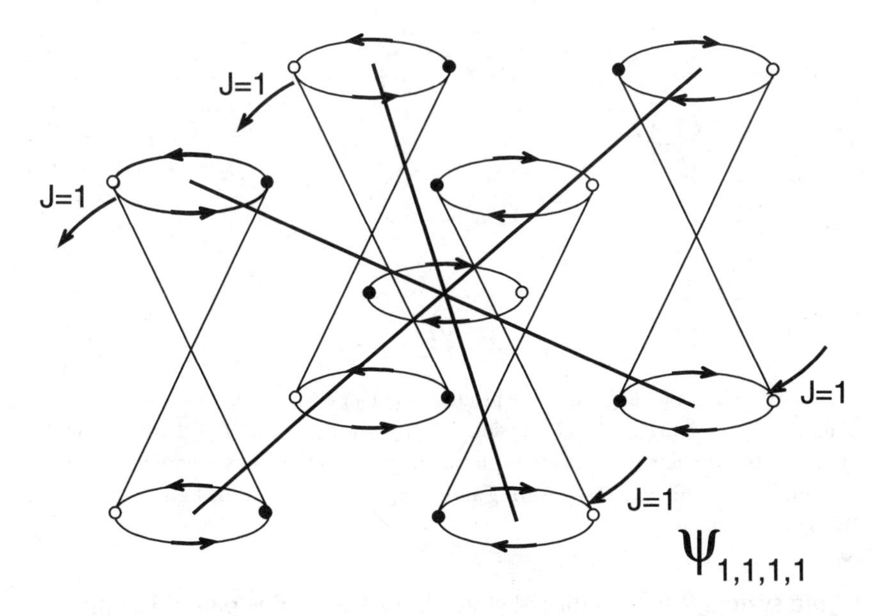

Figure 13.1 Structure of the J/Psi meson, consisting of four sets of two electron-positron pairs, each in a spin 1 rotational state, with the spins of these states opposed in pairs. In this manner, the spin of the whole structure equals that of the central spin 1 neutral pion.

then break up into various other particles of lower mass as seen in the laboratory. Looking through the table of rotational levels in my *Nuovo Cimento* paper, I found that the first excited level of a two pair system had an energy close to a quarter of the J/Psi mass, or 1,495 m_0, close to the mass of the Rho meson of observed mass at the time of 1,519± 6 m_0. Thus, four Rho mesons in a cubic array similar to the hyperon states in shape would give a total value for the mass of 6,076 ± 24 m_0, close to the observed J/Psi mass of 6,060 m_0, or well within the experimental uncertainty.

A spin 1 neutral pi-meson can be strongly bound in the center of the cubic structure. Its mass-energy is canceled by the binding energy of about 34 m_0c^2 in each of the eight bonds holding it to the pions in the corners of the cube. In effect, it is the fulcrum for the rotations of the four Rho mesons that constitute this unusually stable kind of system. as shown in Figure 13.1 above.

Inspection of Figure 13.1 shows that it is possible to obtain the required low value of total spin angular momentum of one unit $\hbar$ for the

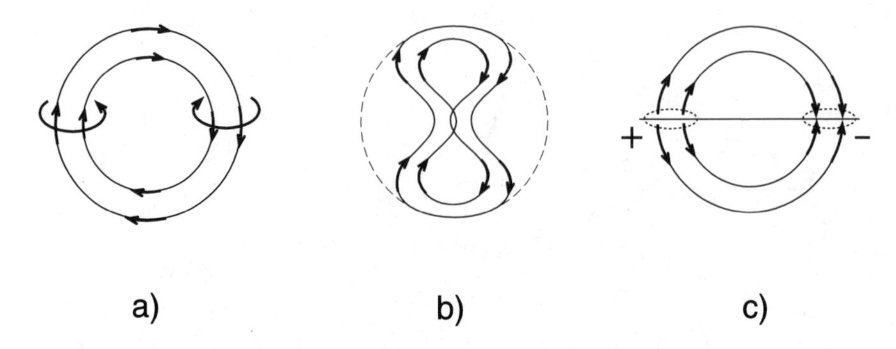

Figure 13.2 Schematic diagram of the production of a pair of charges from a high energy photon or gamma ray, or the conversion of light into matter. (a) Initial vortex ring. (b) Vibration leads to pinching (c) Twisting by 180 degrees leads to lines of vorticity emerging from one region and lines disappearing at the opposite region, or the formation of equal and opposite charges.

whole system if the rotational states have their spins paired in opposite directions, just as in the case of the hyperons discussed earlier, where the planar four pi-meson structures rotate in opposite directions, giving hyperons the same spin $1/2\,\hbar$ for the entire structure, or that of the proton.

This total spin $\hbar$ is required because theory shows that when a high energy electron and positron annihilate each other in a head-on collision, they form a so-called "virtual" photon of spin $\hbar$ which is the first step in a process of successive pair production processes into two, four or more pairs, in exact analogy to pair production by an ordinary photon seen in the laboratory and illustrated in Figure 13.2 above.

The process can be understood in terms of a vortex ring. The initially closed lines of the electrostatic field represent lines of vorticity, around which the ether fluid rotates exactly as does a smoke-ring or tornado closed on itself to form a ring that is normally stable. The first step in the process is a pinching of the vortex ring, when it is caused to vibrate, as for instance from the result of a collision. When oppositely directed lines overlap, it means that the internal rotation of the fluid is canceled, so that the vortex breaks into two half rings. Relative twisting of the two half-rings by 180 degrees leads to the two half rings being held together. In a vortex, the external pressure of the fluid is greater than its internal pressure. The two half-rings behave exactly like suction cups pressed together. Where the half-rings touch each other, there is a cancellation of the internal rotation, ending lines of force. Thus, the cut regions appear to be places where lines of vorticity and thus lines of the electrostatic field either emerge or end. This is what the term "electric charge" means: regions of finite size that are "sources" or "sinks" of electrostatic lines of force.

The lines of electrostatic force are here interpreted as lines of vorticity, around which the ether fluid circulates, just as the air circulates around an imaginary line in the center of a hurricane or tornado. This is a way to interpret the electromagnetic fields postulated by Faraday and put into mathematical form by Maxwell in the nineteenth century, the electric lines of force being the lines of zero internal rotation (like the eye of a hurricane), while the circulating fluid around the line of vorticity would be represented by the magnetic lines of force. In this model, the collision of an electron and a positron would be a rejoining of two halves of a vortex ring to form a temporary or virtual kind of photon, which would then divide to form pairs of massive charges that would in turn form vortex rings and more charges in a series of such processes.

When one accepts the existence of an ether as an ideal fluid in which vortex rings can exist, as shown mathematically by Helmholtz a century and a half ago, with their interactions worked out in detail by Kelvin and J. J. Thomson and others, it is possible to regard "charges" as nothing else than the regions where vortex rings have been cut and "glued" together so that lines of vorticity begin or end in these regions. The rotational motion of fluid circulating around the lines of vorticity or lines of the electromagnetic field represents motional energy. Since Einstein showed that energy is equivalent to mass multiplied by the square of the velocity of light, the regions where these lines exist possess mass or inertia. Thus, it takes the application of a force to set a pair of charges or a "mended" vortex ring into motion, or to stop it once it is moving as a whole unit. It is this property of resisting any change in motion that is called inertia or inertial mass, and it gives all charges "rest-mass," the possession of inertial mass even when not in motion. By contrast, photons do not have such rest-mass, they are always in motion and possess only motional energy. The more compact or the smaller they are the more motional energy they possess. Their energy is related to their diameter d by the relation h/d, where h is Planck's Constant. The diameter d is equivalent to the so-called wavelength of the photon, since when a vortex ring or circular line of the electrostatic field passes by an observer at the speed of light, the pulse recorded by an instrument, such as the vertical component of the field, rises and declines with time like a wave, as illustrated in Figure 13.3 below.

The time taken by the passage of the vortex ring is given by d/c, where c is the speed of light, so that the observer registers a pulse of this duration. The inverse of this time is called the frequency of the wave given the Greek letter ν. Therefore the relation for the energy in a photon E can also be written in the form $E = h\nu$ used by Einstein in his groundbreaking 1905 paper on the discrete or quantized nature of light, a relation that Millikan confirmed with his precision measurements of the photoelectric effect.

Particles of radiation regarded as quantized vortex rings, or trains of such rings moving at the speed of light, behave like finite pulses of waves as they are seen passing by an observer. And since a pair of charges is also nothing but localized motional energy associated with lines of vorticity or electromagnetic lines of force, a passing electron produces a rising and declining field just like a photon. Moreover, both are associated with a local distortion of space, so that they produce a passing gravitational effect, though this may be quite small. In terms of vortices in a fluid ether, the wave-nature of particles arrived at by de Broglie becomes as comprehensible as the particle nature of light arrived at by Einstein.

Possessing mass or inertia is also what gives all types of charged or neutral particles "weight," or the tendency to be attracted by other particles as a result of a gravitational force. But Einstein showed that this mysterious force acting at a distance, according to Newton, is actually due to the distortion of space produced by the presence of matter, regardless of its nature. In the case of a vortex theory, or an electromag-

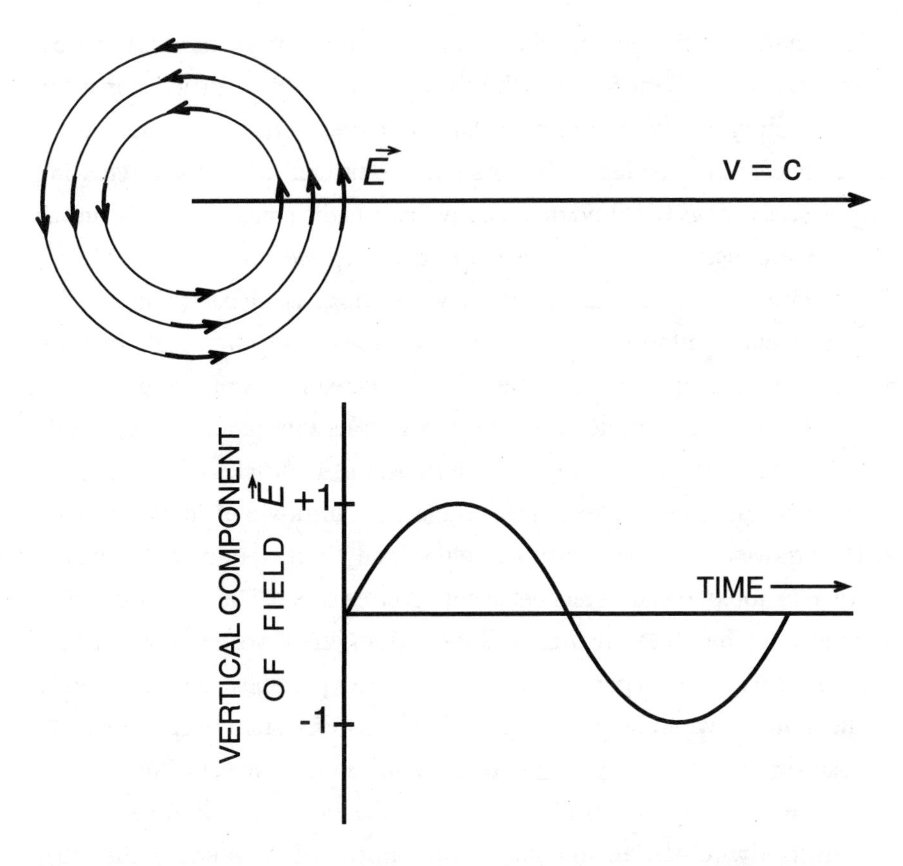

Figure 13.3 Schematic diagram of how a passing vortex ring or circular electromagnetic line of force produces a rising and declining vertical force on a charge that looks like a wave-pulse, having the form of a wave, close to that in form to a so-called sinusoidal wave.

netic theory of matter, mass or localized energy of motion is produced by the rotation of the fluid in every type of vortex, whole or "split and reglued." And where there is vortex motion, space is no longer "flat" or perfectly uniform, just as the liquid surface in a cup of coffee is distorted by small dimples produced by a half-ring of a vorticity when one draws a spoon across its surface, as Helmholtz pointed out at the end of his paper on vortices. It was such a distortion of space by the particles making up the Sun that caused photons from distant stars to be deflected slightly towards the Sun as they grazed its surface during an eclipse in 1919, confirming Einstein's theory on the nature of gravity which he had formulated as the general theory of relativity four years earlier. Thus, once one accepts the idea of a fluid-like ether in which vortices can exist, the mystery of how matter physically manages to distort the surrounding space, never explained by Einstein, is removed. Matter as circulating motional energy of a fluid necessarily distorts the space around it, just as a tornado distorts the otherwise uniform density of air.

With the idea of a fluid ether and vortex rings as the basic entities possessing energy of motion in the region or "field" surrounding them, both light and matter can be understood as localized forms of motion associated with certain amounts of energy and thus inertia and mass. Whole vortex rings, or regular trains of such rings, form particles of light or photons such as gamma rays, x-rays, visible light, ultra-violet light and so on. They move at the maximum possible velocity in the ether, that of light c equal to 300,000 kilometers per second in free space. They are stopped or deflected when they encounter "split and re-glued" vortex rings that are otherwise known as pairs of charges or matter particles, and particularly efficiently so when they happen to hit a massive system of electrons and positrons organized as protons and neutrons in the nuclei of atoms.

Thus, it becomes understandable how powerful photons can be converted into matter, and matter into photons. It happens all the time, when energetic gamma rays coming from outer space strike the particles in our atmosphere. The photons produce electrons and positrons with high energy, in a process called pair-production, and these particles produce photons when they in turn collide with the charges constituting matter, leading to massive showers of particles beginning with a single photon or vortex ring. Similarly, the collision of a highly energetic electron with an equally fast-moving positron in a particle accelerator leads to a photon that immediately divides. That in turn gives rise to an electron and positron in a highly energetic state—the J/Psi—that has the same spin $\hbar$ as a photon. These processes are miniature versions of what apparently happened in the beginning of the universe, when a single photon of enormous energy divided itself to form the first electron-positron pair from which all the particles in the universe originated.

Seen in this manner, it is clear why the discovery of the J/Psi so strongly supported the hypothesis of an electron pair model for the Lemaître atom. In particular, it was the observation that starting with only an electron and a positron, neutrons and anti-neutrons as well as protons and anti-protons were found to be produced in its decay. This is consistent with the fact that the J/Psi possesses exactly the eight electron-positron pairs needed to produce a proton and an anti-proton, each of which requires four such pairs in K-meson states as shown in Figure 12.3 in the previous chapter. But this is precisely what had to happen in the last stage of the Big Bang. Protons had to be produced if a universe as we know it was to come into existence in a phase or structural transition that changed matter from a collection of massive electron pair systems to stable protons.

The electron pair model for the Lemaître atom received still more support from the results of high energy accelerator experiments. In 1977, the discovery of a more massive and thus dense meson was discovered. It was called the Upsilon, and has a sharply defined mass of 18,512.7 ± 0.4 electron masses, or more than ten times that of the proton. As was necessary if these states were to account for the long-lived "seeds" of galaxies and other cosmological systems, it has in fact a still longer lifetime than the J/Psi.

The circumstances that made the J/Psi and the Upsilon possible, namely occurrence of high energy states of the spin 0 and spin 1 relativistic pair at exactly the same energy and also at an energy where a very stable intermediate structure of mesons exists, happen only very rarely. This is why only two such extremely narrowly defined mass-states or long-lived mesons as the J/Psi and the Upsilon have been found. It explains why the Upsilon has almost exactly an integral multiple of the mass of the J/Psi, namely three times its mass, where states of the spin 0 and spin 1 system again happen to coincide.[1]

Besides the fact that an intermediate structure consisting of three J/Psi was located at the observed energy of the Upsilon into which it could transform itself by successive internal pair-production, I found that there was another molecular rotator state composed of neutral pions that has very nearly the same mass-energy. This is a state composed of two sets of high angular momentum K-mesons, each consisting of two electron-positron pairs in the fifth possible state of allowed angular momentum with a resulting mass less than one percent from the observed mass of the Upsilon. Again, with a neutral spin 1 pion between the two rotating pair systems, opposite senses of rotation would lead to the observed spin 1 state of this most massive structure, which was similar to that of the proton, except that in the proton, there was a massive positron at the center.

These models for the J/Psi and Upsilon-mesons not only accounted for the observed masses and spins of these particles, but also for the fact that when high-energy particle collisions reach the point where the Upsilon can be produced, three rather than two jets of mesons begin to emerge occasionally from the point of collision. When these temporary systems disintegrate, they give rise to a large number of ordinary mesons and other particles such as pairs of hyperons, nucleons and anti-nucleon pairs that form jets. Ultimately, all of the unstable particles end up as nothing but electrons, positrons and protons moving close to the speed of light, together with radiation in the form of neutrinos and gamma rays. Thus, the high energy experiments kept supporting the idea that the J/Psi and Upsilon were composed of nothing but relativistic electron pairs.

Other properties of the J/Psi and the heavier Upsilon make them seem consistent with the theory of the primeval atom as excited states of an electron-positron pair that could divide itself in half by a series of "scission" or "fission" processes. It was found that these particles only rarely annihilate into radiation the way the neutral spin 0 pion disintegrates into two gamma rays. The inability to annihilate quickly into two gamma rays was theoretically required for a spin 1 particle, exactly what my model for the J/Psi and the cores of the charged pion and the muon required. As for disintegration into two neutrinos, this was a very slow process by nuclear particle standards. Thus, the fastest way that massive electron pair systems in the very early universe could get rid of their high energy was by dividing successively into two pairs in a series of steps previously seen in ordinary atomic nuclei excited to a high enough energy. In this process, no radiation was emitted, so that these massive electron pair systems could explain some of the so-called dark matter, namely the portion that could not be explained by matter composed of ordinary neutrons and protons, collectively called baryonic matter, in the form of burned-out or collapsed stars and smaller dim objects.

In terms of the relativistic electron pair model for the structural components of the nucleons and mesons that Gell-Mann had given the name quarks, the strings or lines of force associated with the forces between quarks were the highly compressed and confined lines of electrostatic force now regarded as the central lines of miniature tornadoes of very great energy, hundreds of times as energetic or massive as the electromagnetic field associated with electrons at rest. The process of electrons colliding and forming pairs of mesons within a proton could now be visualized as proceeding via the formation of a vortex-ring like a photon or a "gluon" that in turn divides itself into two, each vibrating and quickly and dividing by the pinching process as shown in Figure 13.2 above, then twisting 180 degrees so as to create a "glued together" half-vortex ring composed of two broken half-rings. They hold together as the result of the low pressure that always exists in the interior of vortices such as tornadoes, the higher pressure of the air or fluid circulating at greater distance from the center exerting an inward pressure, just as ordinary photons do. These massive vortex rings that are formed in the bombardment of protons in the high energy accelerators must carry a significant fraction of the excitation energy as they form mesons in successive division processes, explaining the theoretical deduction from the experimental data that gluons and many temporary quark-like particles form within the proton in the course of a high energy collision. At the same time, the stability of the electron-positron systems that form the basic proton structure means that protons are never disintegrated by even the most energetic collisions. The particles found in the jets emerging in a high energy collision are simply the decay products formed as the highly excited electron pair systems constituting the proton return to their lowest or ground-state energy.

The introduction of fractional charges was initially a mathematical device that occurred independently to Gell-Mann and another theorist by the name of Zweig in 1963. In an earlier proposal for three basic constituents of strongly interacting particles published by the Japanese theorists Ikeda, Ohnuki and Ogawa in 1959, namely the neutron, the proton and the Lambda hyperon together with their anti-particles, the charges were either zero or integral multiples of that of the electron. But according to the new proposal of Gell-Mann and Zweig, there were three new entities postulated called "quarks" by Gell-Mann that had charges of one-third and two-thirds of the electronic charge with their anti-particles of opposite charge.. There was the "u" or up-quark with a charge of 2/3 *e*, the "d" or down quark with a charge -1/3 *e*, and an "s" or strange quark also with a charge of -1/3 *e*. Baryons were assumed to consist of three quarks, while mesons were assumed to consist of two quarks so as to always give exactly a charge zero or one electron charge. And now, two more fractionally charged quarks were apparently needed to explain the properties of the J/Psi and the Upsilon as quark-anti-quark pairs, analogous to positronium—with no theoretical reason why many more might not be found.

The development of powerful colliding beam machines together with their very complex particle detectors and the high speed computers that control them made it possible to test the idea that the strongly interacting particles are ultimately composed of electrons and positrons in energetic and compact excited states. The results provided a picture of the nature of the quarks and the structure of the proton that gradually became clearer, although the extreme stability of the proton remained a puzzle for a long time. In the high energy states, the electromagnetic fields of the electron pair system become so highly compressed as a result of their relativistic motion that the forces holding them in equilibrium become the "strong force" of nuclear theory. Because the strength of this attractive force keeps increasing as the energy of the pairs goes up, they are permanently trapped or "confined" inside the stable proton. Therefore, in the electron pair model, the massive quarks that are composed of these electron pair systems are never released in the bombardment of the proton, a fact that cannot be theoretically deduced from Gell-Mann's quark model.

From the discovery of the J/Psi and Upsilon mesons I became convinced that the enormously energetic collisions reached by modern accelerators, equivalent to temperatures of billions of degrees centigrade, illustrated how the universe must have evolved to create matter. Starting with a single energetic electron-positron seed pair at an incredibly high mass, this pair divided until the mass reached that of galaxies, stars, and smaller objects. Ordinary nuclear particles were produced in the last major stage of division, moving in jets of enormous velocity similar to those that had been found to emerge from the centers of active galaxies and quasars discovered since the early 1960s.

But unlike laboratory collisions of high energy electrons and positrons, where protons and anti-protons are always produced in pairs, in the original single electron-positron pair, only one or the other of the two possible spin 1 states shown in Chapter 11 came into existence. Apparently, it was the type that could bind a positron into the stable state rather than an electron. In effect, the universe apparently showed an asymmetry, so that only protons but not anti-protons and neutrons were formed at the Big Bang, rather than equal numbers of anti-protons and anti-neutrons. It was this aspect of the original neutral pion model that allowed the universe to live longer than the fraction of a second during which the newly created particles and anti-particles would have annihilated each other. I began to understand that although advances in technology had allowed us to begin to see aspects of the origin of matter, there was a crucial difference between what could be done in the laboratory even with the most powerful accelerators imaginable and the way the first electron pair was created in the beginning of time. It appeared that the design of matter and the creation of the first pair ensured an asymmetrical universe in which only protons could form and thus survive their birth process, providing the minimum condition for the eventual emergence of life and conscious beings.

But while I was delighted with the discovery of the J/Psi, in the community of theorists it resulted in a state of great confusion. In the seven years since I had begun to work in the application of physics to medicine in 1967 and had stopped being actively involved in particle theory, the idea of just three fractionally charged quarks as the basic constituents of all nuclear particles, as manifested in the abstract symmetry theory of the SU(3) group, had become widely accepted. It was seen as a relatively simple way to understand the nature of the hadrons (the strongly interacting particles), as distinct from the leptons (the electron and muon). As described by Abraham Pais in his 1986 history of modern particle theory, veritable pandemonium broke out in the community of high energy physics as theorists tried to fit this new particle into the existing ideas. The J/Psi apparently required the existence of yet another kind of quark, the "charmed" quark, destroying the simple idea of just three basic constituents of all hadrons.

The discovery of the Upsilon just a few years later required still another type of quark, a fifth fractionally charged particle called the "bottom" quark. This was also not predicted, further defeating the attempt to regard the strongly interacting particles as composed of just three basic entities.

As Pais described it in *Inward Bound,* "the reaction of the theoretical physics community to the [quark] model was generally not benign....The idea that hadrons were made of elementary particles with fractional quantum numbers did seem a bit rich." According to Pais, not since the late nineteenth century, when the reality of atoms was at issue, did the question return: "Is this a mnemonic device or is this physics?" This proposal, although not easy to swallow for most theorists, was a boon to experimentalists who spent the next twenty years searching for such objects, without any firm evidence for free quarks. Other questions that were raised were: How big are quark masses? What forces bind them? Why are they not seen in a free state? How many more kinds of quarks would turn out to be needed? Why were only those quark combinations found that happened to give exactly the charge of the electron or zero?

The main reason why the fractionally charged quark hypothesis was nevertheless gradually accepted by many theorists was that a finite number of fundamental entities seemed preferable to the so-called "bootstrap" hypothesis advanced by other theorists such as Geoffrey Chew at the 1963 Stanford Nucleon Structure conference. According to this hypothesis, the various particles are composed of each other with no preferred "fundamental building blocks" at all, so that in a sense the universe would be envisioned to "pull itself up by its own bootstraps."

Another reason for accepting the existence of Gell-Mann's fractionally charged quarks was an unexpected experimental discovery made at Stanford in 1969. When a new, two-mile long linear accelerator for electrons was completed at Stanford in 1967 producing electrons more than ten times as energetic as the billion electron volts Hofstadter's machine had been able to reach, it was expected that the study of the proton structure would show nothing remarkable. As Pais put it, Hofstadter's experiments were consistent with a smeared-out positive charge distribution and a meson cloud "smooth and soft as jelly" that would not lead to any electrons being scattered by large angles.

Instead, when by 1969 all the instrumentation had been constructed and the new accelerator had been fine-tuned, electrons were found to be scattered strongly over large angles by factors about thirty times greater than had been predicted by theorists. Just as Rutherford's earlier studies of atomic structure using alpha particles at the beginning of the century had disproved Thomson's idea of positive charge being uniformly distributed in atoms, it looked as if there were once again small, point-like entities in the proton that scattered high energy electrons, as if the proton were a box filled with hard nuggets. In view of the great prestige that Gell-Mann had achieved with his prediction of the Omega minus particle discovered in 1964, these point-like entities were widely believed to be evidence for quarks of the three fractionally-charged types.

But then came Richard Feynman to explain the details of the scattering of electrons in a way that the simple three quark model of Gell-Mann could not. In a paper published shortly after those announcing the anomalous scattering of electrons by protons, Feynman suggested that the proton contains a number of long-lived, structureless point-like particles greater than three, each possessing a spin 1/2 and an unspecified charge. Each one of these "partons," as he called them, would carry a certain fraction of the total momentum of the proton like a swarm of free or independent particles as seen from the incoming electron. This electron is then assumed to scatter from an individual parton, and the observed pattern of scattered electrons is explained by summing up such individual parton-electron collisions. The final state of the proton after the collision, which does not contain free quarks, is reached upon redistribution of the shaken-up "box" filled with partons. As Abraham Pais put it, how *that* comes about the model does not tell.

Feynman may have refrained from adopting a definite position on the charges of the particles constituting the proton partly because he knew of my theory that the mesons in the proton were composed of relativistically moving, point-like massive electrons and positrons of spin 1/2 that carried integral charges forming neutral mesons, a result which I had worked out in his office at Cal Tech ten years earlier, and a copy of which I sent him. Moreover, in 1965, the same year that my paper on an electron-positron structure of the charged and excited meson-states appeared in *Nuovo Cimento*, two theoretical physicists by the name of Moo-Young Han and Voichiro Nambu had published a paper in the *Physical Review* in which they suggested that one could in fact have integrally charged quarks. Such quarks would fit the properties of the SU(3) symmetry group if one assumed that each of Gell-Mann's quarks is composed of three different types of particles, in physicist's humor later said to have different "colors" of red, green and blue, even though color is only a property of ordinary matter on the macroscopic scale.

However, as Pais noted in his detailed historical account, the success of Feynman's parton model for the proton raised a difficult question. In the type of scattering of the electrons by the partons assumed by Feynman, "the electron sees the proton as an assembly of *freely moving* constituents. Then why does the nucleon not fall apart?"

It was exactly this puzzle that I had hoped to answer when I went to Stanford in the fall of 1966. During the next seven years I could not pursue its solution, due to my heavy involvement in the problems of medical imaging and the effects of low-level radiation on human health. In fact, I did not become aware of the discovery of the large-angle scattering of electrons at Stanford and Feynman's interpretation of this phenomenon in terms strikingly similar to my approach to the structure of the proton for a long time after these results were published in 1969. As described in my books *Low Level Radiation* and *Secret Fallout,* at that time, in addition to my work in medical imaging, I had become deeply involved in the effort to bring out the danger of nuclear fallout from the proposed nuclear-tipped anti-ballistic missile system and releases from nuclear facilities, based on the disturbing discovery of a large increase not only of childhood leukemia but also of infant mortality from past nuclear weapons testing early that year.

Not until February 1974, when I had finished testifying at hearings on the licensing of a large nuclear plant being built near Pittsburgh, did I feel able to return to the problem of the structure of nuclear particles and their role in the origin of the universe.

Gradually I realized that the great stability of the proton must somehow be connected with a motion of the central positron such that it would always move to maximize the relativistic Coulomb or electrostatic force between the two sets of electron pair systems on the outside. This would allow these pair-systems to rotate freely when struck by another particle, while still being held together by the extremely strong force due to the central charge, just as in the J/Psi. Based on this idea, I was able to arrive at a good estimate of the mass and magnetic moment of the proton, although it would take many more years to modify the model and wait for the results of ongoing experiments to be certain that the proton did indeed have indefinite stability under all conditions as needed for the universe and life as we know it to exist indefinitely into the future.

By 1978, I had satisfied myself that all the latest high energy experiments were consistent with my model that required nine electrons and positrons to form the structure of the proton, and I published a paper to that effect in the *International Journal of Theoretical Physics.* Since the theory of Han and Nambu allowed nine integrally charged "colored" quarks to exist instead of three fractionally charged ones, the electron pair model was consistent with the symmetry-based models as well as Feynman's account of the scattering by partons. It also explained the observed ratio of the number of strongly interacting particles or hadrons created in electron-positron collisions relative to the number of muon pairs if the muon consisted of three charges, rather than the single one of the electron as it did in my model.

However, the extraordinary stability and thus the long life of the proton remained unresolved. Large-scale experiments were being started to see whether the proton might decay spontaneously at an extremely low rate, as had been predicted by Grand Unified Theories, certain new theoretical attempts to unify the various forces. These experiments involved looking at large underground pools of water with sensitive photo-multipliers to detect faint flashes of radiation produced when fast particles were emitted such as would be expected in the decay of a proton. But in the many years of observation since the 1970s, no proton decays were found, so that the lifetime of the proton was greater than predicted by the initial version of the so-called Grand Unified Theory that sought to relate the nuclear forces with the electromagnetic one. Failure to detect a single proton disintegration in the very large numbers of protons in the water of the pools being monitored meant that the proton had to have a lifetime of at least 10^{32} years. This is an enormously large number of 1 followed by thirty-two zeros, a ten billion trillion years, far longer than the presently estimated time since the Big Bang of about 10^{10}, or mere ten thousand million years. Moreover, in the following decades no proton was ever found to disintegrate under bombardment, even though collision energies between beams of protons and other particles reached the enormous energy equivalent to an acceleration to 10^{12} or a trillion volts. The challenge to understand what causes the incredible stability of the proton, now shown to consist of point-like massive particles possessing the spin of the electron kept me pursuing the puzzle of the origin and evolution of matter in the early universe.

CHAPTER 14

THE MASS OF THE UNIVERSE

IN THE DECADE THAT PRECEDED THE DISCOVERY of the J/Psi meson in 1974, cosmology underwent a major revolution. It began in 1963 with the discovery of quasars, star-like objects that turned out to produce as much light as a hundred galaxies, each consisting of a hundred billion stars. These objects were often accompanied by the ejection of jets of particles moving close to the speed of light and the emission of radio waves of enormous power. The inexplicably rapid variation in their power output in just a few weeks or even days indicated that they were extremely compact, dense sources of energy. Suddenly, the universe had revealed itself to be a place of violent phenomena, with the cores of many galaxies showing activity similar to that of quasars. The number of quasars and galaxies that were sources of radio waves was found to increase with distance from our galaxy. The universe was not the same everywhere in space, as the steady-state theory of Bondi, Gold and Hoyle had assumed. Together with the discovery of the cosmic background radiation in 1965, predicted by Gamow in 1948, there was now overwhelming evidence for the Big Bang model developed in the 1920s, strongly supporting Hubble's 1929 conclusion that the redshifts of galaxies were due to their moving away from each other in an expanding, evolving universe.

By 1973, the first orbiting astronomical observatories operating in the ultraviolet region of the spectrum confirmed the existence of heavy hydrogen atoms or deuterium, in concentrations consistent with formation in the Big Bang. But the ordinary nuclear reactions that took place in the formation of the elements could not explain what drove the expansion of the universe against the forces of gravity, nor the incredible power of the quasars and of the active cores of galaxies. Clearly, a previously unknown and extremely powerful physical process was present to account for these mysteries. The explosive division of the primeval atom and its fragments in the centers of newly forming galaxies was such a new process that could have given rise to the hot ball of neutrons that Gamow had envisioned as the beginning of ordinary matter. The primeval atom's ongoing division could also provide the missing physical mechanism for the ejection of dwarf galaxies from the nuclei of large galaxies that Victor Ambartsumian and a few other astronomers such as Halton Arp found in increasing numbers in the 1970s.

Thus, as far as I was concerned, the discovery of the J/Psi in November of 1974 could not have come at a more opportune time. I knew from my work at the Westinghouse Research Laboratory that within the next two decades observational data would be produced by space-based telescopes using electronic imaging techniques vastly more powerful than ordinary photography. The data would indicate whether galaxies are born in the collapse of a gas cloud or from a massive, rotating fragment of Lemaître's primeval atom. Ironically, the secret of the "Old One," as Einstein had put it, was destined to be revealed by the very technology whose development was driven by the discovery of nuclear fission and the construction of nuclear weapons that Einstein regretted so deeply.

Early in 1974, I had decided to ask the new head of the Department of Radiology, Dr. Ralph Heinz, to have my associate Donald Sashin take over the day-to-day operation of our research group to allow me to return to the problem of nuclear particles and astronomy. This was the aspect of my work that I had had to put aside for the past seven years. Heinz agreed to my proposal when I explained to him that developments in the instrumentation for nuclear physics and astronomy could have important applications in medical imaging. I spent the next few months in the library, catching up on all the latest developments in high energy physics and astronomy.

Popular literature was full of the changes in the field as well. In May, 1974, *National Geographic* spotlighted recent developments that would enable us to probe the secrets of the universe. I had been approached by the author of the article, Kenneth Weaver, for information on recent developments in electronic imaging techniques in astronomy, so I was particularly interested in the story. Among the technical advances the author mentioned was the Copernicus Orbiting Astronomical Observatory, which was gathering previously inaccessible spectrographic data in the ultraviolet. The article also discussed the new electronic imaging system guiding the Mount Palomar telescope, linked to a computer that intensified light by a hundredfold. Both instruments used the SEC Vidicon that we had developed at Westinghouse. As the astronomer Allen Sandage, who carried on the study of distant galaxies begun by Hubble, explained to the reporter, it was now possible to get a spectrum in nine minutes that had taken 42 hours back in the 1920s. The new electronic imaging technology allowed astronomers to probe farther and farther into space to learn how fast it was expanding and whether that expansion was slowing down.

After World War II, observers had found many objects that emitted radio frequencies but were invisible in ordinary light. Some of these sources were found to coincide with nebulae that were gas and dust clouds, others with remnants of supernovae or exploding stars, and still others with galaxies. But until 1960 nothing as small as a star was known to be such a radio source.

Things changed radically when Allan Sandage and the radio astronomer Thomas Mathews discovered an unusually faint star-like object in the exact location of the radio source whose designation was 3C48 (number 48 in the *Third Cambridge Catalogue of Radio Sources*). These strange objects were of great interest to astronomers all over the world. In the next few months, other radio sources were identified with optical objects that emitted strongly in the ultraviolet as seen by the orbiting observatories, as well as in the long wave radio end of the spectrum, fluctuating in intensity over short periods of time. They were named "quasi-stellar radio sources" or quasars for short.

Early in 1963, at the California Institute of Technology in Pasadena,

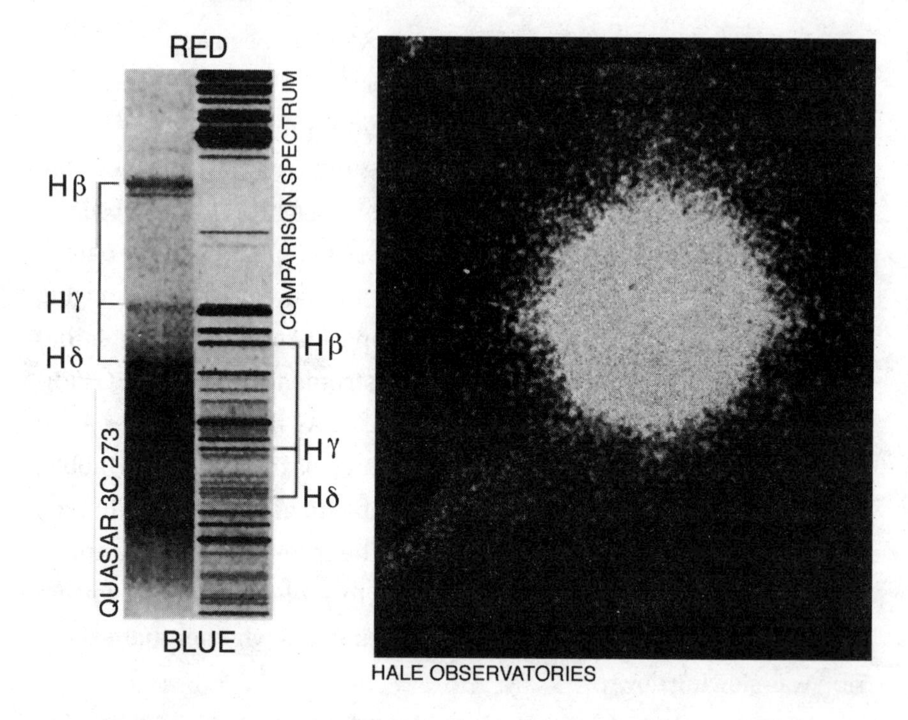

Figure 14.1 Optical image of quasar 3C 273 on the right, and its spectrum on the left, showing the characteristic three lines produced by hydrogen shifted upwards towards the top or red end, compared with a reference spectrum.

astronomer Maarten Schmidt examined the spectrum of a newly discovered quasar, 3C 273 (shown in Figure 14.1). He was troubled by the fact that the emission lines in the spectrum were not familiar.

Spectra clearly identify chemical elements. One set of lines, for example, is easily recognizable at certain wavelengths as indicating emissions from excited hydrogen atoms, as first described theoretically by Bohr back in 1913, while other patterns identify heavier elements such as silicon, calcium, iron and other elements of greater mass. But, the spectral lines of 3C 273 did not fit any known element.

As Kenneth Weaver told it, "Suddenly it dawned on Dr. Schmidt that the familiar pattern was there, but not in the expected place. It was shifted far down towards toward the red, or longer wavelength, end of the spectrum. This red shift was much larger than those astronomers were accustomed to seeing. 'That night,' the astronomer recalls,' I went home in a state of disbelief and said to my wife 'Something really incredible happened to me today.'"

Astronomers were soon forced to believe that the spectrum of 3C 273 was indeed shifted to the red by a large amount when many more such objects were found. In fact the next quasar, 3C 48, was determined to have a red shift twice as large as that of 3C 273, corresponding to velocities that indicated distances of billions of light years, the distance light traveling at 186,000 miles per second covers in the course of a year. And since the universe was estimated to have a present size on the order of some ten to fifteen billion light years, the quasars were apparently the most distant objects. However, some astronomers believed that quasars were not at these extreme distances because this assumption required such enormous amounts of power.

I believed that the quasars were at cosmological distances since the explosive decay of fragments of the Lemaître atom, which had a mass of the entire universe could explain the incredible rate of energy output by a highly compact object that ordinary nuclear fusion processes could not explain. I had become convinced that the quasars represented "mini-Bangs." It seemed most likely to me that they were somehow delayed examples of the creation process.

Maarten Schmidt shared the majority view that quasars were indeed at huge distances from us, and like me, felt that they were probably the nucleus of a newly forming galaxy. Indeed, there was growing evidence that massive objects were at the center of many if not most galaxies, which in a Lemaître-type of cosmogony would be the remnants of still more massive original seed pairs, the descendants of the electron pair with which the universe would have begun.

Just exactly how the quasars are produced, what is the source of their energy, and how the energy is channeled into jets of material, as can be faintly seen in the lower left-hand corner of the photograph of 3C 273 above, would be the subject of many years of research, debate and controversy in the scientific community. But when the J/Psi was discovered, I felt it was overwhelming evidence that Lemaître's ideas on the origin of the universe, the galaxies and stars were likely correct. It became clear that I would have to work out the details of the processes that would answer the many questions raised by these parallel developments on the microscopic and cosmic scale . And I had to accept the idea that this would be a very long task, since I could only work on the problem part-time. At the end of 1974, I had to return to my mounting work in radiological imaging. I also worked on the continuing campaign against the nuclear arms race by publishing studies of the health effects of past nuclear testing.

There were now two major problems in particle physics and cosmology that were linked, in my mind. On the one hand, I had to find a way to test the idea of the initial decay of the Lemaître atom as a massive electron pair system that would divide repeatedly and give rise to the observed structures in the universe, and on the other hand, I had to understand how stable protons would be formed from massive electron pairs in the last explosive stage of the division process Lemaître had envisioned. Moreover, the model of the proton would have to explain not only its size, mass and stability, but also the internal distribution of charge and source of its magnetic field that Hofstadter had discovered. All had to tie in the properties of the point-like partons and quarks, which constituted the amazingly stable structure of the proton.

Thus, while educating myself on the latest developments in astronomy and cosmology, I also had to learn of all the new results in high energy particle physics. The first thing I did was to study the results of the electron-positron collider experiments that had led to the discovery of the J/Psi. As mentioned earlier, the J/Psi allowed me to account for the ratio of the production of pions to muons, with nine electrons and positrons in the proton and three electrons and positrons forming the muon, giving a reasonable estimate of the proton's mass and size. Once I had succeeded in getting an article on the electron-positron structure of the proton published in *The International Journal of Theoretical Physics* in 1978, I had hoped to be able to turn my attention to the question of the evolution of the early universe.

Unfortunately, this work was interrupted the following year, in March of 1979. The Three Mile Island nuclear reactor had a serious accident, on the second day of which I was asked to fly to Harrisburg on behalf of environmental organizations deeply concerned that the public was not being told the full extent of the danger. There I joined the well-known Harvard biologist George Wald to answer questions from the media, at which time we recommended the evacuation of pregnant women and young children, who were known to be most vulnerable to the effects of the radioactive gases. In the morning, as I was about to leave for Harrisburg, my mother called me from her home in Buffalo, New York, to tell me that she felt ill. I urged her to have a neighbor take her to the hospital right away, and told her that I would call my brother in New York and ask him to fly to Buffalo immediately and said that I would come to see her as soon as possible from Harrisburg in the afternoon. When I called the hospital from the airport in Pittsburgh where I had to change planes shortly after noon on my way to Buffalo, I learned that she had just died of a sudden massive rupture of the aorta.

The next weeks and months were some of the most difficult in my life, as I subsequently described in *Secret Fallout: Low Level Radiation from Hiroshima to Three Mile Island.* Moreover, my wife had just accepted a position at Indiana University in Bloomington, Indiana, so that we had to sell the home we had built, and I had to arrange to divide my time between my work in Pittsburgh and a new part-time position as Adjunct Professor in the Department of History and Philosophy of Science in Bloomington.

It took the better part of two years before I was able to spend any significant time on my work in astronomy and particle physics. When I resumed my reading in these areas, I found two articles that gave me a way to quantify the mass of the initial relativistic electron pair, or the highest energy state to which such a system could be excited. One of these was by the Russian astrophysicist V. L. Ginzburg, written in 1971, which discussed the problem of the initial state of the universe. Although general relativity seemed to suggest that the initial state was a point-like and therefore of infinite density, a so-called "singularity," quantum theory seemed to require that in the actual world, there existed an upper limit to the possible density of matter given by Planck's Constant *h* divided by 2π or $\hbar$, the speed of light *c*, and Newton's gravitational constant *G* in the simple form $c^5/\hbar G^2$. It was the only way these constants could be combined to give a mass divided by a volume, and this so-called Planck Density had the enormous value of about 5×10^{93} grams per cubic centimeter, that is 5, with 93 zeros before the decimal place.

A second article that had been published in 1975 by J. A. Wheeler and C. M. Patton suggested that at this unimaginably enormous density, space tears itself apart, producing holes or pairs of charges. Since my model of the Lemaître atom consisted of such a pair of charges, I asked myself what volume would be required to produce the Planck Density if the mass of this pair was indeed that of the known universe. How would its size compare with that of my relativistic electron pair?

In 1979, three astronomers, R. P. Kirshner, A. Oemler and P. L. Schecter, published a new estimate of the present average density of visible matter in the universe, from which I was able to get a rough estimate of the mass of the universe out to the present radius to which it was believed to have expanded by now, the so-called Hubble radius, estimated at ten to fifteen billion light years. The mass of the visible universe worked out to roughly 10^{80} protons, corresponding to 1.7×10^{56} grams. The initial volume occupied by this mass divided by the Planck Density gives the extremely small volume of 3.4×10^{-38} cubic centimeters, a very small number consisting of 3.40 preceded by 38 zeros between the decimal place and 3.4. To get a rough idea of what the size of the primeval atom given by the Planck Density was, I simply took the cube root of this volume in my pocket calculator, which turned out to be 3.2×10^{-13} centimeters. To my amazement, this was extremely close to the classical diameter of the electron of 2.8×10^{-13} centimeters, some four times larger than the orbit of my electron-positron pair system, or about the volume into which most of the electromagnetic field-energy of the relativistically contracted first pair was packed.

It was a crucial result that confirmed my belief that Lemaître's idea would eventually prove to be the key to the nature of matter and the structure of the universe in purely electromagnetic terms. No other known form of matter could achieve such an enormous mass in such a small volume. Yet it was the finiteness of this volume that avoided the problem of the infinitely dense singularity which Einstein's General Theory of Relativity by itself necessarily led to, as had been shown by the British mathematicians Stephen Hawking and Roger Penrose in the early 1970s. Without my planning for it, the effort to work out the structure of the neutron in Paris back in 1957-58 led to the discovery of the existence of a minimum approach distance between any two relativistically moving charges, following a method of calculation proposed by Einstein in his 1905 paper on special relativity. And now the work on the model for the neutral pion, composed of an electron and a positron rather than a proton, that Feynman had made me work out in his office in 1960, turned out to automatically overcome the problem of the singularity at the beginning of the universe that seemed to be required by Einstein's 1915 theory of general relativity.

In effect, in 1960 I had stumbled upon a way to connect general relativity with quantum theory, since the rapidly precessing orbits of my quantized electron pair system led to a non-Euclidean geometry for an observer in the extremely rapidly rotating reference frame in which the orbits were at rest. Thus, the primeval atom of Lemaître, assumed to be an enormously energetic, rotating and thus massive electron-positron pair, had to produce a universe whose geometry was of a closed, non-Euclidean type. This non-Euclidean geometry and the mathematical techniques to deal with it had been developed in the nineteenth century by Riemann and other mathematicians, and they were used by Einstein to develop his locally curved space explanation of gravity, followed a few years later by his model for a static closed universe.

The electron pair model for the initial state of the universe had another great inherent advantage. It allowed one to understand the physical meaning of a closed space in terms of the electromagnetic field

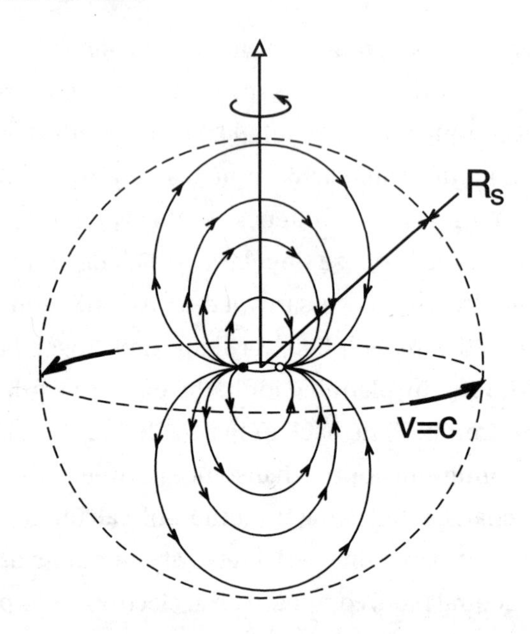

Figure 14.2 Schematic diagram of the electric lines of force associated with a pair of equal and opposite charges, assumed to rotate like a rigid structure with the outermost region moving at the speed of light.

lines of force invented by Faraday, which begin at a positive charge and end on a negative one, moving outward to large distances but eventually curving back towards the center where the pair of charges is located, as shown in Figure 14.2.

As Faraday visualized it, it is the lines of force that represent physical space, and he saw electromagnetic radiation as vibrations traveling along these lines. Thus, small oscillation of a charge relative to another would send waves or pulses traveling outward along the lines of force, but eventually they would return to the center. As a result, no light or radiation of any form could travel outward to infinite distances, so that the space would be a closed one. If one now assumes that the charges and their associated fields rotate like a rigid structure as the work of Gödel, as corrected by Ozsvath and Schücking in 1962, required for a rotating Einstein universe, and if lines of force can never move faster than the speed of light as the system rotates, then there will exist a maximum radius to which the lines of force of a newly created electron pair can extend. This is the radius at which the tangential velocity of the spherical field structure will have reached the speed of light. It corresponds to the so-called Schwarzschild radius that in general relativity defines the maximum radius which a light signal can reach, separating one space from another.

Using Faraday's idea of a system of lines of force produced by two equal and opposite charges, together with that of a rigid rotation, one can answer the age-old question: What lies beyond the edge of a closed universe? It is nothing other than the same fluid-like ether that is assumed to exist within the closed universe, except that it contains no lines of vorticity or circulation and thus no motion. It is this motion that represents energy and, by Einstein's relation $E = m c^2$, it is this energy of motion that gives rise to its mass. And since no waves or motion in the lines of the electromagnetic field or vorticity can ever move beyond the point where the lines of force reach, no energy is ever lost from the universe. It must therefore contain exactly the amount with which it was created, and will contain it indefinitely into the future.

It therefore seems that Maxwell's theory of electromagnetism, which requires an upper limit to the velocity of field lines as developed mathematically on the basis of Faraday's experiments and physical concepts, leads naturally to a positively curved Riemannian space of the type that Einstein had arrived at. And when the space-time substrate is regarded as an ideal fluid ether, the lines of electric force can be regarded as the center-lines of tubes of rotation or vorticity around which the ether rotates, like an invisible miniature tornado representing motional energy pervading all of physical space.

Although we hardly think of our world in this way, according to this model for matter, we live in a rotating "black hole" from which no light can escape. Because we don't find ourselves crushed out of existence, and since relativity leads to the existence of a minimum approach distance, it is clear that there are no "singularities" at the center of black holes, "naked" or otherwise in the actual world composed of nothing but electrons and positrons. Thus, although black holes also exist in an electron pair theory of matter, they are nothing more unusual than regions where there is a sufficient density of matter within a finite radius such that, according to the theory of relativity, the gravitational force prevents light or even the most rapidly moving particles from escaping.

As in the case of a closed universe, the average density of matter in a large black hole can be very small, and will be in the range observed for normal matter. I realized that even when a very massive star collapses under the action of gravity due to the pressure produced by its nuclear reactions no longer being capable of countering it, its matter does not collapse into a singularity of infinite density, but merely down to about the same density that exists in ordinary atomic nuclei. The reason is that the centrifugal force within each electron pair system constituting the nucleons always exactly balances the attractive force. As a result, adjacent pairs always maintain a minimum distance from each other corresponding to the smallest possible approach between any two of the members of the pair-system. This is the same minimum distance between electrons and positrons that explains the observed masses in the molecular structures that make up stable protons as well as unstable mesons.

The forces that fix the structure of the proton are balanced so that it can never be crushed out of existence, even by the greatest compressive forces that can be imagined in the most massive of collapsed stars. Black holes created in the collapse of burned out stars much more massive than the Sun undoubtedly exist, but according to the electron pair model of matter, they will end up with essentially the same maximum density as ordinary protons or neutrons, and thus comparable to the density of so-called neutron stars. This is of course very large, about 10^{14} grams per cubic centimeter, or 100 trillion times the density of water (which is 1 gram per cubic centimeter), but it is far less than the Planck Density of the first pair, and certainly not infinite.

The simple electron pair model of matter therefore eliminates the subject of many science fiction stories, namely what happens inside a black hole. Unfortunately for science fiction writers, it also eliminates any possibility of using them to travel rapidly to other parts of the universe, or even to other universes, through "wormholes," as well as any speculations about travel backward in time. There simply are no such things in the electron pair model. All the many ways theoreticians have speculated about them apply only to a purely classical theory of general relativity in which the nature of the ultimate particles of matter was left undefined, as I had learned from my conversation with Einstein. General relativity by itself had nothing to say about the nature of the ultimate constituents of matter, and it continued to be widely regarded as unrelated to electrodynamics and even irrelevant to quantum and particle theory.

Indeed, in the best account of what was known about the early universe in the late 1970s, Weinberg's *The First Three Minutes,* the author had discussed the problem of the universe's very early state when a large variety of particles were believed to exist in thermal equilibrium with radiation. As he put it, "we simply do not know enough about the physics of the elementary particles to be able to calculate the properties of such a mélange with any confidence. Thus our ignorance of microscopic physics stands as a veil, obscuring our view of the very beginning."

The evidence suggested to me, then and now, that the electron and positron are the ultimate stable entities with which the universe began. They are the only truly indivisible elementary particles that have been observed in the laboratory since their discovery in 1897. I realized that if indeed the electron pair theory of matter turned out to be correct, it would represent an enormous simplification of our ideas about the very early universe, in which no other fundamental entities and no singularities would be found.

The electron pair model eliminates another speculative idea proposed by a few theorists and also adopted in popular science fiction scenarios, namely that some day a spontaneous quantum fluctuation might give rise to another universe that would have different physical constants and that might therefore destroy our present universe in its explosive birth process. Since it takes a large amount of energy to create any relativistically rotating electron-positron pair such as the pi-meson, the J/Psi and the Upsilon, it must have taken an enormous energy to create the first pair. Thus, a relativistically rotating electron pair system for the origin of the universe cannot have arisen without a huge net energy input, exactly as is required in the case of the large accelerators that produce the J/Psi and Upsilon in highly energetic collisions. There is therefore no possibility of a future small quantum fluctuation giving rise to another universe in the electron pair theory of matter that could threaten our existence. Nor is there the possibility of an advanced civilization some day creating one "in the basement" with little net energy input, as has been suggested by some theorists. In the electron pair model, there is no free lunch, as some theorists have jokingly put it, so that the Old One apparently cannot avoid a week of hard work.

Still another concept advanced by a few theorists loses its attraction in a model where rotational motion provides permanent stability. This is the idea of many baby universes existing besides our own. In the absence of a singularity at the beginning, both the laws of nature and the energy to produce the first pair of a universe must have existed indefinitely in the past and will continue to so in the future. But both a finite or an infinite number of universes existing together would collapse into a single one under the action of gravity according to both Newton's and Einstein's theories of gravity, just as would a finite or an infinite collection of stars. This has most recently been discussed by Guth in *The Inflationary Universe*, who points out that in this case, Newton had made a mistake in believing that if the stars were uniformly distributed throughout an infinite space, there would be no center towards which they could collapse.

Although Einstein had laid the foundations of modern cosmology, he had not been able to accept the idea of a rotating universe that the mathematician Kurt Gödel had suggested to him. It would have required the concept of an absolute reference frame or absolute substance such as an ether before the universe had begun, relative to which the concept of rotation could be defined. Apparently, the early influence of the philosopher Ernst Mach, who did not believe in such an absolute reference frame, was too strong. Thus, instead of the well-known centrifugal force produced by rotation relative to the absolute reference frame of Newton, Einstein was forced to introduce an unknown type of repulsive "anti-gravity force" in the form of the so-called cosmological constant in order to keep a non-rotating closed universe precisely stable for an infinite time.

Einstein eventually adopted the idea of some sort of space-time continuum for the medium that was distorted by the presence of matter in his theory of general relativity, but he could not bring himself to accept it as an absolute reference frame in which rotation could provide the anti-gravity force he needed to achieve his hope for a stable universe. Nor could he accept the concept of an ideal frictionless fluid in which Helmholtz had shown that vortices could exist indefinitely in a stable form, which seemed to provide a possible hydrodynamic model for his photons and the quantized charges of electromagnetic theory. Einstein was unable to fully adopt the idea of a purely electromagnetic field nature of the electron's mass without any "ponderable" matter of possibly point-like size as used in quantum field theory, a subject that he and I had discussed at some length in the course of our conversation back in 1947. However, it seemed that in this case, Descartes' ideas about vortices in the ether and Newton's belief in an absolute reference frame would turn out to be vindicated.

As a further test of the finite-sized electron pair model for the Lemaître atom, I decided to calculate the mass of the universe theoretically, based on the ideas of Dirac. In the late 1920s Dirac had developed the relativistic wave equation that automatically led to the existence of electron spin and predicted the existence of the positron. In 1937, Dirac had drawn renewed attention to the surprisingly simple relations between certain large numbers that occurred in cosmology. For example, the ratio of the size of the observable universe to the size of nuclear particles is of the order of 10^{39}, which is of the same order of magnitude as the ratio of the electromagnetic to the gravitational force between protons of mass M_p, simply given by e^2 / M_p^2G, where G is Newton's gravitational constant. Furthermore, the square of this ratio, namely about 10^{79}, is of the order of the number of protons estimated to be in the visible universe. Dirac argued that these and other simple relationships involving cosmological quantities were unlikely to be pure coincidences, and that somehow these relations had to be explained in terms of a model for the evolution of an expanding universe. He suggested that they might be explained if the gravitational constant varied with the time of evolution, declining slowly from its initial value. Other theoreticians, such as Robert Dicke at Princeton University and Carl Friederich von Weizsäcker in Germany also considered a possible variation of *G* in the course of evolution of the universe. But theoretical arguments by Edward Teller, based on what the effect of a varying *G* would be on the temperature of the Sun and thus of the Earth a few billion years ago, argued against it. Moreover, recent precision astronomical measurements carried out since these theoretical ideas were proposed also seemed to rule out a variation of the strength of universal gravity with time.

As a result, the origin of Dirac's large number relation had remained a mystery. And so I decided to see what these numbers would give for the mass of the universe if the basic particles were the electron and positron rather than the proton and anti-proton. If Dirac's original idea was correct, then the square of the ratio of the electrostatic force between a positron and an electron should give the mass of the universe in units of the electron mass m_0. This is the ratio e^2 / r^2 to m_0^2G / r^2, or simply e^2 / m_0^2G. Inserting numerical values into my hand calculator, I found this ratio to be 4.167×10^{42}, whose square is 1.736×10^{85} in units of the electron mass m_0. Since the proton has a mass of 1836 m_0, one finds that in units of the proton mass, that of the universe calculated according to Dirac's relation gives a theoretical value of 9.455×10^{81} proton masses or almost 10^{82}. But this is a factor about one hundred times larger than the visible mass of 10^{80} protons as estimated by Kirshner and his colleagues, consistent with the evidence that only about one percent of the mass of the universe is in visible form.

Thus the electron pair model had passed still another test, in that it gave a reasonable value for the total mass of the universe of about 10^{82} proton masses or 1.736 x 10^{85} electron masses, and about ninety-nine percent of this total in the form of dark matter. This result was in accordance with the latest observations that included the dark halos around galaxies found by Rubin and Ford in the 1970s, resulting in an average total density of all forms of matter for the universe as a whole very near the critical one needed to close the universe. The model appeared to lead to the kind of stable, closed universe that Einstein had originally postulated, a universe that would neither fly apart to infinity or collapse into a point-like singularity in which all life would come to a fiery end.

Moreover, the model had solved the mystery of the origin of the Dirac Large Number Relation, since it is based on the hypothesis that the electron and not the proton is the fundamental building block of all matter, with an integral and not a fractional charge. Therefore, in this model all matter had to be composed of charged electrons and positrons, including the quarks in the nucleons and mesons, as Han and Nambu's integrally charged quark theory and recent Grand Unified Theories implied. The assumption that electrons and positrons were the only true elementary particles explained the masses, spins and life-times for the muon the pion, and the neutron, and accounted for all the high energy resonance state seen so far. All these results convinced me that there would be no need to postulate some as-yet-unknown form of dark matter particles producing new types of forces that were being widely discussed in the literature as perhaps constituting the invisible or "missing" matter in the universe.

Much of the dark matter could turn out to simply consist of old, burned out stars, neutron stars, stellar black holes, planet-sized or other, smaller objects such as meteors, comets, and still lesser fragments of the original primeval atom ejected to large distances in the explosive "mini-Bang" that had to accompany the formation of every cosmological structure in a Lemaître-type model. And the existence of quasars and active cores of galaxies over a wide ranges of distances indicated that there were apparently delayed mini-Bangs in which new galaxies were created, as Maarten Schmidt had conjectured, together with vast amounts of gas and dust ejected into intergalactic space.

All these results strongly suggested that some of the original fragments resulting from the Lemaître atom had managed to survive in the massive centers of large galaxies for very long time periods. A fraction of these massive electron pair fragments were apparently ejected long after the Big Bang, as the Russian astrophysicist Novikov and the Israeli physicist Ne'eman had in fact suggested independently in the mid-1960s. The remaining clusters of massive primordial pairs would still be located in the centers of the largest cosmological structures greater than superclusters of galaxies. Their nuclei would be so massive that they would be invisible black holes, yet they could account for a dominant fraction of the total mass of the universe even today. Indeed, the famous French scientist Laplace, who was one of the first to contemplate the idea of black holes, had envisioned just such a possibility at the beginning of the nineteenth century.

In any case, for me, the failure to find any of the speculative new kinds of long-lived particles that theorists had recently dreamed up—such as supersymmetric particles, axions and magnetic monopoles that they hoped might explain the dark matter despite decades of high-energy experiments—strongly supported the simple electron pair model of matter. The clear prediction of the model that no other type of nuclear particle must ever be found, other than those that would end-up as electrons and positrons, would remain a crucial test of the theory. As beautifully simple as it was, it was also a very vulnerable theory on which I had placed all my bets, something I could not have done if I had not had a "cobbler's job" all my life, as Einstein had advised me.

Another test of the electron pair theory involved the ultimate or limiting radius of the expanding universe that the model predicted. It had to be much larger than the presently observed universe determined from its current rate of expansion as calculated from the Hubble constant, since there appeared to be as yet no sign of any definite slowing-down. Moreover, it would have to give a result for the time since the Big Bang when protons and the first elements like helium were formed, that was at least as great as what astrophysicists calculated to be the age of the oldest stars.

This had been a problem in the years immediately after World War II, when astronomical observations of the redshift and the estimated distances of the galaxies had given a time since the expansion began shorter than the age of the Earth, of about 4.5 billion years. This had been one of the reasons why Bondi, Gold and Hoyle had proposed the steady-state theory, according to which there was no Big Bang at all. This was an error that began to be corrected by the German-American astronomer Walter Baade only in 1952. Baade used the one hundred-inch Hooker telescope on Mount Wilson near Los Angeles, during the blacked-out nights of World War II between 1941 and 1945. He took advantage of the darkness to resolve individual variable stars in the central bulge of the nearby Andromeda galaxy for the first time. He discovered that there were two different types, reddish ones in the bulge and bluish ones in the arms. The latter were much brighter, and from these Hubble had underestimated the distance. By the 1970s, some studies had indicated that there were still greater errors in the distance estimates, so that by the early 1980s the age of the universe had grown to about ten billion years. This was more than twice the age of the earth and closer to the twelve to fifteen billion years calculated for the oldest stars. Thus, the final, limiting value for the radius of the expanding universe, or the Schwarzschild radius given by the electron pair model, would be another crucial test.

The simplest test was to calculate the Schwarzschild radius from the theory of general relativity using the best observed value of Newton's constant G and the mass of the universe M_u given by the theory using Dirac's relation, 1.736×10^{85} electron masses, equal to 1.703×10^{58} grams. This maximum radius of the universe in Einstein's theory, named after the German astronomer Karl Schwarzschild who first exactly solved the equations of general relativity, is simply $2GM_u/c^2$, which works out as close to 2.5 trillion light years. This is some two hundred times greater that the recent estimates calculated from measurements of the Hubble constant, or ten to fifteen billion light years, so it turned out to be consistent with what was in fact observed. If the universe has been expanding at the same steady rate since the birth of ordinary matter without slowing down, then the time that has elapsed since the Big Bang has been ten to fifteen billion years. But such a constant increase in size with age is exactly what the action of the centrifugal force balancing the pull of gravity would lead to, preventing the slowing-down predicted by Friedmann's theory for a non-rotating universe.

Ironically, this counter-gravity action of centrifugal force produces the equivalent action of the cosmological constant Einstein had to introduce into his equations to obtain a stable closed universe, but now such a force opposing gravity resolved the puzzle why the expansion of the universe was not being slowed down as fast as Friedmann's model predicted. Thus, the rotating electron pair model gives a value for the ultimate size that is much larger than that deduced from the redshift of the galaxies assuming no rotation, resulting in an age of the universe since the birth of ordinary matter consistent with the age of the oldest stars, removing the problem of a universe younger than the stars it contains. Thus, by 1981 when I had a chance to get back to the problem of the initial state of the universe, the model had passed a very crucial test.

According to this rotating model, the universe is very far from having reached its maximum size, consistent with the observational evidence that currently there does not appear to be any definite evidence for a slowing-down of the expansion rate due to gravity. This means that space appears to be nearly Euclidean or "flat," with a positive curvature corresponding to the closed, finite size that it will attain only in another two trillion years or so. It also means that there is no need to postulate an extremely rapid "inflation" or rapidly accelerating expansion of the universe by many orders of magnitude from an infinitely dense state in a small fraction of a second at a speed many times that of light.

Indeed, I realized that the rate of expansion before the final step of decay of the Lemaître atom could have been a very slowly accelerating one, as the original pair-system divided in a series of steps, leading to a period of accelerating growth of physical space. Such a relatively slow rate of expansion, at less than the speed of light, was similar to the case

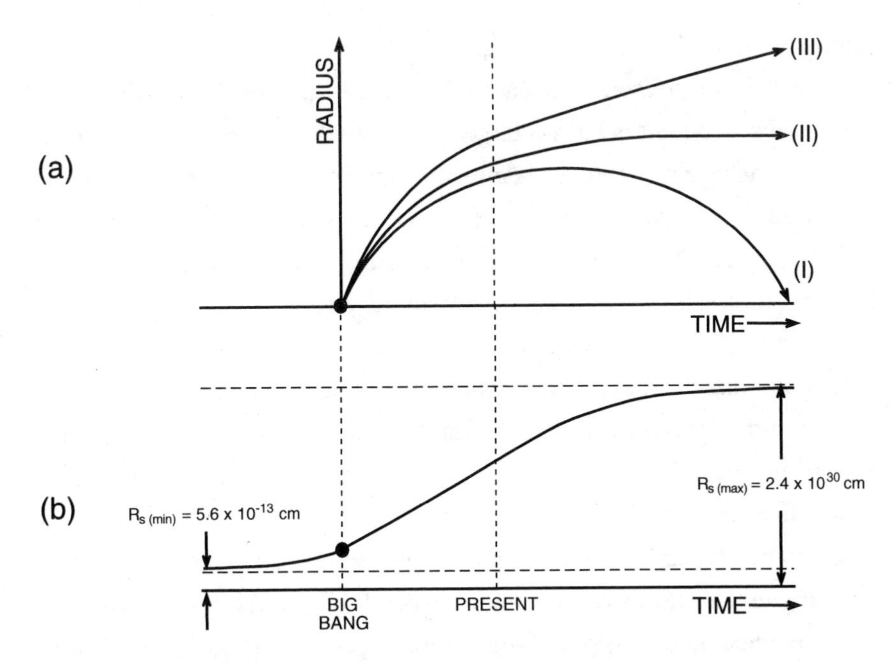

Figure 14.3 Schematic diagram illustrating the expansion of the universe according to various models. Figure (a) shows the three different cases possible in the non-rotating Friedmann case: Curve (I) is the case where the universe starts from zero size with too small a motional energy and collapses again. Curve (II) is the critically balanced case and Curve (III) is the continually expanding case. Figure (b) represents the rotating pair model that starts from a single electron-positron pair of finite size, accelerates as the pair divides repeatedly until the moment when neutrons and protons are formed, giving rise to the Big Bang, after which the expansion proceeds at a steady rate where gravitational forces are balanced by centrifugal forces, until it approaches a stable limiting radius.

examined by the Dutch scientist Willem de Sitter shortly after Einstein had published his General Theory of Relativity in 1915. In the electron pair model, this gradual acceleration would end with the formation of nucleons and turn into a steady rate of expansion until the maximum possible radius dictated by Einstein's theory would be reached, the Schwarzschild radius, as illustrated schematically in Figure 14.3 above.

Two different types of scenarios are illustrated in Figure 14.3. In Figure 14.3(a), the case of the non-rotating Friedmann model is shown for three different cases, each starting from zero size, or an infinitely dense singularity. In all three cases, the initial expansion rate is very rapid and slows down under the action of gravity.

Curve (I) represents the case where the initial motional or outward directed kinetic energy is not large enough to overcome the pull of gravity, and the universe reaches a maximum radius like a stone thrown into the air at less than the "escape velocity" of 24,000 miles per second on Earth, so that the universe stops expanding and falls back towards the center, ending its existence in a fiery singularity.

Curve (III) is the case of more than enough initial motional energy in the radial direction to overcome the pull of gravity, causing the universe to expand to an infinite size, so that the galaxies become increasingly separated with time, with their stars gradually burning out and life "ending with a whimper."

Curve (II) is the so-called "critical case" where the energy with which the universe expands is exactly that needed to keep it expanding forever at a decreasing rate, approaching a limiting state that is highly unstable since there is no centrifugal force to balance the pull of gravity. The problem connected with this particular scenario is that it requires an extraordinary degree of "fine-tuning" of the initial condition and a perfect uniformity in all directions to prevent the universe either from collapsing or flying apart to infinity. The limiting size is the case of Einstein's original static universe, for which he had to introduce an arbitrary "cosmological constant" representing an anti-gravity force of unknown origin to keep it in equilibrium, something that he later regretted, but which now appears to be the equivalent of a centrifugal force, leading to the case represented in Figure 14.3(b).

In Figure 14.3(b), the case for a rotating and expanding universe is shown schematically, where the expansion starts from a single relativistically rotating electron-positron pair of finite dimensions so that there is no problem of an infinitely dense state where the laws of physics break down. As a result of successive divisions by two, the size of the universe or the space occupied by the individual pairs of decreasing mass, there is a long period of accelerating growth or "inflation" as in the original solution of Einstein's equations found by de Sitter and revived on an extremely short time-scale by Guth. This period of accelerating growth comes to an end when the neutrons and protons are formed with the liberation of a great deal of energy, more than needed to produce the nucleons from the massive electron pairs in the last stage of the division process in a so-called phase transition. This extra energy goes into the acceleration of the newly formed protons and neutrons, heating them to a temperature of about a trillion degrees, thus creating the hot ball of neutrons Gamow discussed as the initial state of matter back in 1946-48 when I was a part-time teaching assistant and graduate student sitting in his lectures. This was the moment when an extremely hot sphere of neutrons existed briefly before it expanded and cooled, just long enough for the formation of helium and a few other nuclei heavier than the proton in a gigantic fusion reaction. It was an enormous hydrogen bomb-type of explosion which, together with the heat produced by the formation of the neutrons, became the source of the three degree cosmic background radiation in the form of radio waves that Penzias and Wilson discovered in 1965.

After this brief period of neutron and proton formation, followed in a matter of minutes by the production of helium and other light elements, the cosmos kept expanding at a steady rate under the action of gravity balanced by the combined force of the initial explosion and the centrifugal force due to rotation. According to this model, in some trillions of years in the future, when the outward directed motional energy has spent itself, the universe will finally achieve the stable state that Einstein postulated. But such a permanent or unchanging universe, as had been widely accepted for centuries, only comes into existence as a result of the centrifugal force that was impossible for Einstein to consider. This is because rotation is a meaningful concept only relative to distant matter, as the positivist scientist Ernst Mach had maintained. But in the beginning of an expanding universe, when there were no distant fixed stars, rotation had meaning only relative to a pre-existing, infinitely extended ether that could serve as an absolute reference frame—what Newton regarded as the body of God. The religious conception of the nature of the universe was a concept that the young Einstein, even as a boy in high school, rejected, according to his biographer Philipp Frank. In later years, he somewhat modified his views.

The broad picture of how the universe originated when a massive electron-positron pair divided to form the seeds of galaxies, stars and ordinary matter was finally beginning to come together and seemed to be supported by the various tests that I had been able to carry out so far. But much more needed to be worked out in detail and compared with observations, such as the origin of the various cosmological structures in the early universe as well as their masses and sizes. Nor had I solved the riddle of the stability of the proton and its formation in the Big Bang. I had not even attempted to calculate the temperature that the early universe would reach on the basis of this model, which was crucial to explain the observed cosmic background radiation and the initial formation of helium and other elements. This was the task that I began to undertake when a fortunate set of circumstances allowed me another semester of reduced commitments to other obligations, this time as a result of a sabbatical leave granted to my wife in January of 1983.

CHAPTER 15

THE BIRTH OF EVERYTHING

LONG WALKS ALONG THE BEACH in front of our apartment just thirty miles south of Cape Canaveral, where the exploration of space had begun, allowed me to examine what I had learned about the origin of the universe, and what I still needed to understand. There was now overwhelming evidence for evolution from a highly compact state—most simply understandable as a single high mass electron-positron system from which protons and all other forms of matter had originated. The masses and lifetimes of all the newly discovered nuclear particles and the strength of the forces between them, could be explained in terms of such a basic pair of particles, with only the input of the four fundamental atomic constants. That argued for the kind of underlying unity Einstein had believed in all his life.

The next step was to learn what aspect of the division process led to the large-scale organization of the universe into systems of galaxies, stellar clusters and stars in the kind of hierarchical organization of rotating systems that Kant had postulated and a few astronomers like de Vaucouleurs believed in.

Lemaître had estimated that about 260 steps of division by two would be involved in the process. The story goes that on a five-day trip across the Atlantic, the imaginative British astronomer Sir Arthur Eddington (with whom Lemaître had studied for a year), had calculated by hand that this would produce just about the number of protons needed to account for the visible mass of stars in the universe, the number two multiplied by itself two hundred and sixty times, or 2^{260}, producing close to the number ten multiplied by itself seventy-eight times, or 10^{78} protons. Now I had concluded that if, instead of protons, the ultimate entities were electrons, and the large number hypothesis of Dirac therefore had its origin in the number of electrons in the universe, there would be 1.736×10^{85} electron masses in a finite, closed universe. And so I asked myself: exactly how many steps of division by two would it take for the initial system to divide itself down to the lowest possible state that would produce a pair of pions flying apart in opposite directions, so as to conserve linear momentum in the process?

With an electronic hand calculator not available to Eddington in the 1930s, this was a very simple question to answer. The average theoretical mass of a pion taken from my 1961 paper, corrected for an error in the mass by one electron-mass that I discovered since it had appeared, was readily calculated to be given in terms of the fine-structure constant a as having the value $(2/\alpha) - 2$ electron masses.[1]

The result for the number of divisions was so hard to believe that I repeated the calculation again and again, and each time the display gave the same answer: the number of divisions by two was exactly twice the inverse of the fine-structure constant $2/\alpha$.

This meant that based on the model of the neutral pion that I had worked out in Feynman's office, and whose mass was determined by the fine-structure constant, the mass of the universe was given by the same, pure, dimensionless number $2/\alpha$ that had determined the mass of the neutral pi-meson as a relativistic Bohr-type electron-positron system. The mass of the universe could be written most simply as $[(2/\alpha) - 2]\, 2^{2/\alpha}$ times the mass of two electrons.

The odds against this result occurring by pure chance were close to nil. But the implications were amazing. It meant that the Dirac large numbers were not independent quantities, but determined by a single number, the only pure number that could be derived from the four basic atomic constants governing electrodynamics and quantum theory. And since the square root of the number of electrons in the universe equaled the ratio of the electromagnetic to the gravitational force between an electron and a positron, it meant that not only the mass of the universe but also Newton's constant of gravitation was determined by the fine-structure constant within the experimental uncertainty. The theoretical value was 6.6721 x 10^{-8} compared with the observed value of 6.6720 ± 0.0040 x 10^{-8} in centimeter-gram-seconds units.[2]

I had found a connection between gravity, electromagnetism and quantum theory, coming up with a value for the gravitational constant in agreement with the most recent measurements within 0.0015 percent, compared with the forty-times greater experimental uncertainty in the observed value of Newton's gravitational constant of ± 0.060 percent. A single pure number seemed to determine the nature of our world, all the way from that of the nuclear particles, atoms and molecules to stars, galaxies and the universe as a whole, a design that was extremely unlikely to have come about by chance. As Einstein had put it so well, "I cannot believe that God plays dice with the universe." And it was a beautifully elegant design, evidently simple enough for humans to understand and admire it rather than to find it hopelessly complex. As Einstein also liked to say, "God is clever, but He is not malicious."

The time had come to write a paper about these findings, and by the end of our stay in Florida in April 1983, I had finished it and sent it to *Nature* in London, which had published an article of mine on the discrete energy losses of electrons in solids in the late 1950s. But the paper was rejected by the editor, John Maddox, as too speculative.

I was at least in good company. The same thing had happened to Fermi with his ground-breaking paper on the theory of the emission of neutrinos and electrons from radioactive nuclei in 1934. And so, I followed his example and submitted a shorter version to an Italian journal, *Letters to Nuovo Cimento*, where it was accepted and published in October of 1984.

In this article, I discussed another important implication of the idea that the universe evolved in a series of divisions by two from a single electron pair. If the process of creating new charges is to continue beyond the first step, then the Planck Density with which the process began had to decline with each division. The reason is that the volume in which most of the field-energy is concentrated is constant: eight times the inverse of the fine-structure constant times the cube of the classical electron radius or $(8/\alpha)(e^2/2m_0c^2)^3$. This is very close to the volume of a non-Euclidean sphere, which has a radius equal to two classical electron diameters, $2\pi^2(2e^2/m_0c^2)^3$, a quantity that remains constant throughout the division process because the fundamental atomic constants involved must not change with advancing time in the electron pair model. This radius is analogous to the mean radius of the volume in which the electron's field energy is concentrated according to quantum field theory. But if all the atomic constants do not vary throughout the history of the universe, in accordance with astronomical observation of distant and thus old galaxies and stars since the Big Bang, as Gamow had pointed out in 1968, then the only way the Planck Density $c^5/\hbar G^2$ can decline with decreasing mass is if *G* increases at every division.

This was startling, although a change in *G* as time passes had been considered by Dirac as well as other theoreticians. But they had postulated a decrease as the age of the universe increased. However, after much thought, I realized that an increase in the strength of gravity was only a localized change in the curvature of space in Einstein's theory, and that the value of *G* for the universe as a whole could remain constant at the value determined by the mass of the first pair. In that case, as the local strength of the space-curvature force increased with every step, it would reach a value equal to the Coulomb Force when the mass was that of an electron, or sufficient to hold the electron together in the face of the strong repulsion between its field-lines at their source.

Einstein had considered a local gravitational or space-curvature force strong enough to keep the electron stable in a paper he presented at a meeting of the Prussian Academy of Science in 1919. (He and I had discussed this question at length when we met in Princeton in 1947.) In fact, Lloyd Motz had worked out a model for the electron based on exactly such a large local strength of the gravitational force in his 1962 paper. In my 1961 paper on the pi-meson, I had noted that the dependence of the mass on the distance between the two charges had the same mathematical form as for two masses in general relativity. This was equivalent to a non-Euclidean geometry for the extremely rapidly rotating pair-system, or the existence of a very small Schwarzschild radius in which the world of the electron pair was enclosed.

Here was an inner, small "curled-up" dimension of the kind Theodor Kaluza and Oskar Klein had discussed back in the early 1920s, in an effort to arrive at a theory that would unify electromagnetism and gravity. The possibility of such an inner small space associated with the fundamental particles of matter was also an idea about which I had begun to read in accounts of superstring theory. A locally high space-curvature or gravitational force was not quite as strange as it had at first seemed.

Such a locally confined region of high spatial curvature was also related to the idea of a vortex of a size roughly equal to the volume into which the field energy of the electron pair was squeezed or confined. Here was a possible connection between the ideas of Helmholtz, Maxwell, Kelvin and others about "vortex atoms" in the nineteenth century and the ideas of a finite, stable classical electron as worked out by Motz. The concepts of a small "curled-up" dimension, and matter as vortex atoms, are also closely related to the idea of matter particles as composed of superstrings.

In the 1980s, superstring theory proposed to replace infinitely small point-particles with tubes of small but finite size as the fundamental entities. Without any dynamic space-time models for nuclear particles in terms of electrons and positrons, the diameter of the superstrings was assumed to be twenty orders of magnitude, or ten million trillion times smaller than the electron or any other nuclear particle.[3] But the growth in the strength of the local gravitational constant, along with decreasing particle mass that the electron pair theory requires, would allow the superstrings to become comparable in diameter to the electron and proton. What is currently the highly abstract theory of superstrings would then apply to the field or lines of vorticity between relativistically rotating electron pairs, and come into contact with the actual world of ordinary particles.

Superstrings are thought to be composed of nothing but the space-time continuum of the General Theory of Relativity, in the electromagnetic theory of mass, is the fluid ether. Closed strings can therefore be identified with the vortices that constitute the quanta of radiation, and cut, or open, strings as the electromagnetic fields between electrons and positrons.

The local increase in the value of G also decreases the so-called Planck Mass M_{Pl}, given by $(\hbar c/G)^{1/2}$ whose magnitude has the enormous value of 2.176×10^{-5} gram, the equivalent of 10^{19} protons. A Planck Mass in a volume equal to the cube of the Planck Length is the maximum possible density in quantum theory, or the Planck Density $c^5 / \hbar G^2$. This is the density of the initial state of the universe in the Lemaître theory, composed of an electron-positron pair, or of a vortex ring of the mass of a quantum of light from which it may have originated.

The Planck Mass, or the Planck Energy given by $M_{Pl}\, c^2$, with its extremely high value of about 10^{22} million electron volts, also plays a crucial role in the various GUT or Grand Unified Theories. The Planck Energy is where the strength of the weak nuclear force associated with the neutrinos, the strong nuclear force associated with the nucleons, and the electromagnetic forces acting between electrons and positrons are assumed to take on the same strength and thus become unified. But, with the high local value of G for the electron and positron pair system, this unification already happens at 285 million electron volts, close to the mass-energy contained in two pi-mesons. In the electron pair theory of mesons and nuclear forces, electrons and positrons moving with this energy are able to account for the strength of the nuclear force and the lifetime for the emission of neutrinos associated with the weak force. Present GUT theories require enormous energies to reach the point that forces become unified—unobtainable in experiments, since they require linear accelerators light-years in length. According to the electron pair theory, the energies necessary for reaching the "holy grail" of Grand Unified Theories have long been achieved in existing machines.

A local rise in the strength of the gravitational constant, as the primeval atom divided into lower mass seeds of cosmological systems, also means that the Schwarzschild radius, which defines the size of the closed space, or the outer limit to which the rotating electromagnetic field can reach, would not decline quite as rapidly as it would with declining mass if the value of *G* were constant. Instead, it turns out that with a growing local *G*, the same relationship between mass and size exists for every cosmological system as it does for the universe as a whole, as if each system has the same basic characteristics except for its scale. For every system, the mass is proportional to the square of the radius, as if the mass were a plane disk of constant density, exactly the kind of system into which old spiral galaxies collapse. But that is also the relation between mass and size of large vortices such as hurricanes and tornadoes, and also of the vortices that Descartes visualized, in which planets were carried around a central star. Particles of matter and light as vortices in a fluid ether thus receives still further support.

Indeed, a number of astrophysicists, for instance Martin Rees at Cambridge University and Menas Kafatos at George Mason University, have pointed out that the large cosmological structures, from superclusters to galaxies and stellar associations, all have exactly this relation between their mass and their size as the pair-theory requires, namely a mass that increases as the square and not the cube of the radius, as illustrated in the graph of Figure 15.1 below.

Large cosmological systems fit a straight line on this type of logarithmic plot very well. When extended downward, the plot reaches the size and mass of an electron pair, exactly what a vortex model for these charges predicts. Given the mass of a cosmological system, the model will correctly predict its size or spacing. Alternatively, given the size of a large cosmological structure, it predicts its mass, greatly strengthening the hypothesis that all the systems in the universe evolved from a single electron pair of enormous mass, and that the local value of the gravitational constant increases in proportion to the square-root of the mass. The very compact astronomical objects such as stars and planets have their mass correctly given by the line but their observed size is smaller than predicted due to their vastly greater density.

Following our stay in Florida, I had a chance to discuss these results with friends and colleagues in different parts of the country. In August of

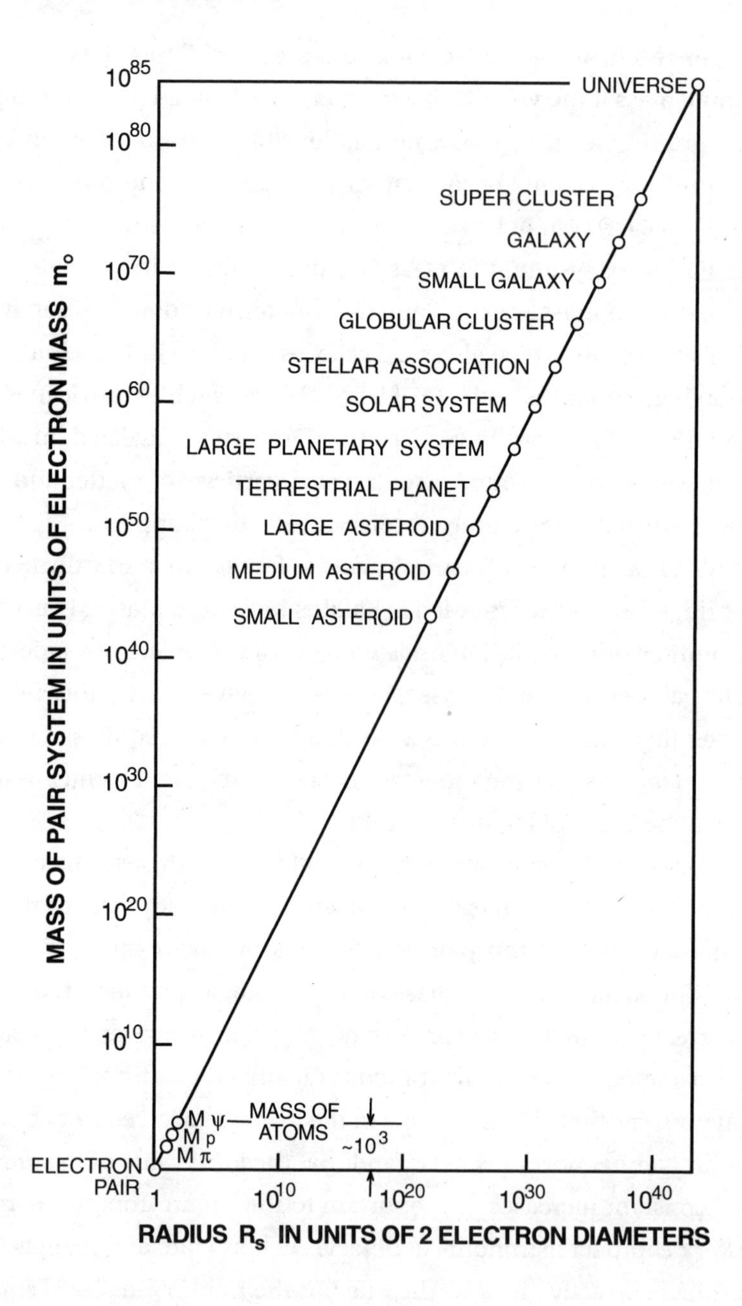

Figure 15.1 The relation between the mass and size of cosmological systems required by the electron pair model is shown as a diagonal line. Only the observed masses and sizes of the large, diffuse systems fit this line. For the dense, compact astronomical objects, the observed masses fit, but the observed radii are smaller.

1983 I visited Edwin Salpeter in Ithaca, a German-born physicist like myself who had been a young research-associate when I was a graduate student at Cornell, and who had since developed a theory for the formation of the heavy elements in stars, as well as a theory for the energy source of black holes in quasars. At the Carnegie Institution in Washington, D.C., I discussed the idea of rotating cosmological structures with my former classmate at Cornell, Vera Rubin, who had been the first to suggest that the universe of galaxies appeared to be rotating, and who, together with her colleague Kent Ford, subsequently discovered the existence of massive dark halos around rotating spiral galaxies. Surprisingly, she told me that the rotating superclusters she and de Vaucouleurs had proposed in the 1950s were still a possibility, and that there were studies underway to see whether the large streaming motions of groups of galaxies then being reported involved rotation. In Boston, I discussed the basic ideas of the theory with my thesis adviser Philip Morrison, who had moved from Cornell to the Massachusetts Institute of Technology, and who had by then become a widely known lecturer and book reviewer for the *Scientific American*. As usual, he asked some tough questions which gave me further food for thought, but I felt encouraged by his interest in my ideas.

Back in Bloomington, Indiana, I had gotten to know the astrophysicist Stuart Mufson who invited me to give a seminar in the astronomy department. In the course of the seminar, I pointed out the regularity in the ratio of the masses of the system, roughly spaced by a factor of a thousand, and with typical numbers of members ranging from about 500 to 2,000 in each of the systems, such as the number of galaxies in a supercluster or the number of stars in a stellar association. It began to look as if there were major stages in the divisions by two of the primeval atom, and that a portion of the massive pre-baryonic pairs were delayed in their decay. These portions apparently survived as the seeds or nuclei of the structures, providing a strong gravitational force which caused a fraction of the initially ejected matter to re-collapse. In the case of galaxies, this process would form an initial spherical or elliptical type, or the central bulge of spiral galaxies, while the arms might represent a delayed ejection, as Ambartsumian had suggested. I kept pursuing this idea since it seemed to explain both the initial explosive quasar phase of a newly evolving galaxy and its subsequent evolution from a spherical shape towards extended spiral and irregular disk forms. This was the direction of evolution originally suggested by Eddington and adopted by Hubble as a means of classifying galaxies, although by the 1950s, the evolutionary sequence was believed to be in the other direction, from a widely diffuse to a compact elliptical or spherical form. When I was asked what observable predictions my model makes that could be tested, I suggested that looking at the very distant and thus earliest forms of galaxies would be most important. They should show a preponderance of compact spherical or elliptical galaxies, with an increasing number of spiral and diffuse disk shapes as they evolve and get closer to our epoch.

In November, I presented my first paper on the Lemaître model at a meeting of the Physical Society in San Francisco. I took advantage of the occasion to visit Luis Alvarez at Berkeley, a high energy physicist who had developed the first large liquid bubble-chamber for recording nuclear particle events more efficiently than could be done with cloud chambers or photographic emulsions. He had tried to persuade me to come to California to work on imaging systems and I had gotten to know him quite well over the years. He asked me the one question for which I still did not have an answer, namely how I could explain the great stability of the proton. It was to be another six years before I found a possible answer.

I also visited Hofstadter at Stanford. I had not seen him in the years that I had to put aside all my work in particle physics. By then, he was developing new techniques of imaging the arteries in the heart using the X-rays from a large circular electron accelerator at Stanford, work very closely related to ours at the University of Pittsburgh. We shared a wide range of interests in instrumentation for medicine and space research, solid state physics and the nature of the fundamental particles of matter, and we remained friends who would occasionally meet until 1990, when he died..

In January of 1984, I had the first opportunity to visit de Vaucouleurs at the University of Texas in Austin and to discuss my ideas about the origin and structure of the universe. Born in France in 1918, he had moved to the U.S. in 1957 with his astronomer wife and frequent collaborator Antoinette, largely concentrating on the properties of galaxies beginning in 1960. One of the earliest proponents of the existence of superclusters, in the 1970s de Vaucouleurs developed a view of the universe as a hierarchical organization which differed from the widely accepted view that the galaxies were randomly distributed with a constant density in space. By the time I saw him, evidence was beginning to be found that there were indeed large-scale structures packed with clusters of galaxies, vindicating his ideas. He had known Lemaître personally and was clearly sympathetic to my ideas, listening with interest, and commented that he had not followed up on his early work that indicated a rotation of the local supercluster. He was particularly interested in the possibility of deriving a value for the gravitational constant from the four atomic constants and wrote my theoretical value in the corner of the blackboard for future reference. I felt extremely encouraged by this visit with one of the world's outstanding authorities on the nature and organizations of the galaxies, and it helped me to persist in my research.

By the time I met de Vaucouleurs again, at the June 1992 meeting of the Astronomical Society, I had worked out an explanation for the origin and energy source of quasars and active galaxies. I had first presented these ideas the previous fall in a paper presented at a meeting on the nature of active galactic nuclei at the University of Maryland and published in the proceedings by the American Institute of Physics. De Vaucouleurs and I found some chairs and sat down together in front of my poster-paper displayed on a large bulletin board, and I began to explain what I had learned about the masses and sizes of the cosmological structures since we had first met in Austin.

I told him that I had been guided by the following facts. The first was that the masses of the various systems seemed on average to differ by a factor of a thousand. Secondly, I observed that the members of the larger systems also numbered about a thousand, which was to be expected on the basis of the kind of repeated division process of an original primeval atom. Thirdly, I had realized that the volume where most of the field-energy or mass of the original electron pair systems was located was just large enough to accommodate about a thousand electron-positron pairs, packed as closely as the minimum distance between them would permit. In this region of space—where most of the field-energy or mass of the seed pairs were located—the field-strength was large enough to tear holes in space, as Wheeler and Patton had phrased it, creating pairs of charges of decreasing mass by a repetitive division by two.

This was the inner space or dimension defined by the size of the fundamental atomic constants. With its high local gravitational curvature, it was the inner, "curled-up" dimension of Kaluza-Klein's theory and of the new superstring theory that could on average accommodate about $8/\alpha$ or 1,096 pairs. The idea occurred to me that perhaps after every ten divisions by two of the primeval pair, which would result in 1,024 pairs each with a mass of 1/1,024 of the first pair, this inner space would be filled up. Then, as the next division by two was beginning to take place, there would no longer be room for all the pairs, and the majority would be forced out. This would leave a shell of some twenty percent (or some two hundred) of the pairs held together temporarily by the immense gravitational force that acts between them. It can be shown mathematically that the electron pairs in the center of a spherically symmetric distribution do not experience a net gravitational force, so that physical processes are not slowed down and they are able to divide as if they were in free space, unlike the pairs in the outer shell.

The pairs that were pushed out of the critical volume in which they were created in turn divided ten times, and the process of expulsion was repeated twenty-seven times. A tightly packed, rapidly growing embryonic hierarchical system resulted, in which the whole future organization of the cosmos was laid down on a microscopic scale, just as in the case of a human embryo. A simple calculation shows that even by the time the seeds of all the visible stars had been formed in the eighth major stage of ten divisions by two, all the 10^{23} "seeds" of stars that had formed in the division process could have been packed into a tiny cube about a few millionths of a centimeter on edge. At one time, it was truly Blake's "world in a grain of sand."

This theory explains the early origin of large-scale structures that could not have formed from the random collapse of a diffuse gas in the relatively short time since the Big Bang. Nor could the very large systems of galaxies have formed so soon after the Big Bang without massive seeds having been laid down before it took place.

A discovery announced shortly before the June 1992 meeting of the Astronomical Society supports the idea that the seeds of cosmological structures already existed in the early universe. High precision measurements of the cosmic background radiation made by the COBE satellite showed small but definite non-uniformity that had to be produced before the universe was a mere 300,000 years old. This was the time when the universe had cooled to about 3,000 degrees centigrade above the absolute zero of -273 degree centigrade. At the relatively low temperature of 3,000 degrees, compared to the millions of degrees during initial element formation in the Big Bang, the electrons could be captured by the protons and the nuclei of greater mass into the weakly bound, large atomic orbits to form ordinary hydrogen, helium and a few other heavier elements in the gas and dust filling space. Low energy photons could then begin to travel through the gas without colliding with free electrons and escape, forming the nearly uniform background radiation photons whose wavelengths were stretched by the expansion of space until they became detectable as radio waves. Thus, the original radiation generated in the initial explosion had first taken on the statistical equilibrium form of that emitted by an ideal black body at very high temperature, which when it decreased to about 3,000 degrees radiated a thermal spectrum of the same shape—detected in the form of radio noise in the millimeter and centimeter range by the COBE satellite.

The slight non-uniformity in temperature across the sky—at a level of only one part in a hundred thousand—that the sensitive COBE satellite detected was therefore hailed as proof that the early fluctuations of density required to seed the formation of galaxies and larger structures existed, as George Smoot and his colleagues reported at the April 1992 meeting of the American Physical Society in Washington.

When I met with de Vaucouleurs two months later, I placed a copy of the COBE satellite's picture of the non-uniformity in the cosmic background radiation on the display board, since it also supported the existence of early structure formation according to my model. I pointed out to de Vaucouleurs that it was the problem of accounting for the surprisingly early formation of galaxies, superclusters and still larger structures that had caused theoreticians to postulate some form of cold-dark matter that would provide density fluctuations around which these structures could form rapidly enough. But no candidates for such a form of matter had been found. There were hopes that perhaps neutrinos might not be pure forms of radiation without any rest-mass so that they might have a small finite mass, much less even than the electron, and thus serve as seeds for structure formation. This was called "hot-dark matter," because low mass neutrinos would move rapidly, in contrast to the much more massive "cold-dark matter particles" proposed by particle theorists. Scientists speculated that statistical fluctuations early on in the evolution of the universe might have been the seeds of the voids and walls of galaxies that were observed by the redshift surveys of the distribution of galaxies in deep space. But none of the proposed causes of structure formation seemed to work for both small and large distances at the same time. Moreover, all the theoretical proposals implied a purely random evolution of structures in the cosmos, whereas there were increasing findings of regularity in the large-scale patterns of galaxy distribution, as a hierarchical model of the type produced in the decay of the Lemaître atom required.

In 1990, the astronomer T. J. Broadhurst and his colleagues had published the results of a deep survey of a thin, pencil-like region of space in an article in *Nature*. They had found a surprising regularity in the distances between superclusters, with an average value of 463 million light-years. As I pointed out to de Vaucouleurs, the spacing of the superclusters in this deep survey were close to what would be expected theoretically on the basis of the electron pair model for the initial spherical ejection of objects into space. This theoretical value was close to three times the distance between the centers of superclusters after gravitational collapse into closely packed disk-like arrangements, where the factor three was simply the square-root of the cube-root of 0.8 times 1,024, the average distance between superclusters initially ejected from a central cluster of massive seeds at the time of the Big Bang and uniformly distributed in a spherical form. Three times the theoretical value of a supercluster diameter of 151 million light-years was 453 million light-years, very close to the observed spacing. The more or less circular voids seen in the regions nearer to us, as reported by the Harvard astronomers Margaret Geller and John Huchra the year before, and which I believed to be collapsed, rotating superclusters, had an average diameter of about 140 million light-years, as close to the theoretical value as the uncertainty in the value of the Hubble Constant would give. Thus, the size of circular voids with their dense walls fitted amazingly well to the size of superclusters that arose in the model as the third stage of ten divisions by two of the Lemaître atom. The centrifugal force, due to the rotation, together with the initial force of the explosive formation of nucleons, apparently pushed most of the galaxies to the rims of the superclusters. Their observed sizes were also close to that for the local supercluster that de Vaucouleurs had arrived at in the late 1950s

As shown in Table I, the sizes of the smaller structures in the universe also show diameters or distances between centers in agreement with a highly organized universe. The average spacing of galaxies such as ours (the system next smaller in size to superclusters), is 2^5 or 32 times smaller, namely 1.45 megaparsec or 4.7 million light-years using the fact that one parsec equals 3.261 light years, in reasonable agreement with the value of 5.06 million light years obtained by P. J. Peebles in 1989 . For the case of stellar systems, the theoretical prediction for the average spacing of 1.38 parsecs or 4.5 light years fell between the measured distances to the three nearest stars to the Sun, Proxima Centauri, 4.27 light-years; Alpha Centauri, 4.37 light-years; and Bernard's Star, 5.9 light-years.

Table I

Masses, sizes and rotational periods of cosmological systems predicted by the electron pair model of matter.

STAGE n_s	SYSTEM M_O*	MASS IN SOLAR MASSES	SPACING ($2R_S$) OR DIAMETER IN MEGAPARSECS#	ROTATIONAL PERIOD ($2\pi R_S/c$) YEARS** DAYS HRS SECONDS
0	Universe	7.95×10^{24}	1.52×10^{6}	1.56×10^{13}
1	Supercomplex	7.76×10^{21}	4.75×10^{4}	4.87×10^{11}
2	Complex	7.58×10^{18}	1.48×10^{3}	1.52×10^{10}
3	Supercluster	7.40×10^{15}	4.64×10^{1}	4.76×10^{8}
4	Large Galaxy	7.23×10^{12}	1.45×10^{0}	1.49×10^{7}
5	Dwarf Galaxy	7.06×10^{9}	4.53×10^{-2}	4.65×10^{5}
6	Globular Cluster	6.90×10^{6}	1.42×10^{-3}	1.45×10^{4}
7	Stellar Association	6.73×10^{3}	4.42×10^{-5}	4.54×10^{2}
8	Stellar System	6.58×10^{0}	1.38×10^{-6}	1.42×10^{1}
9	Large Planetary System	6.42×10^{-3}	4.32×10^{-8}	4.43×10^{-1} = 162 days
10	Small Planetary System	6.27×10^{-6}	1.35×10^{-9}	1.39×10^{-2} = 5 days
11	Large Asteroid System	6.12×10^{-9}	4.21×10^{-11}	4.33×10^{-4} = 3.8 hrs
12	Small Asteroid System	5.98×10^{-12}	1.32×10^{-12}	1.35×10^{-5} = 427 sec
-		——	——	——
-		——	——	——
27	Proton Formation	4.19×10^{-57} ##	***	$9.62 \times 10^{-26} = 1.1 \times 10^{-20}$ s.

*1 Solar mass equals 1.989×10^{33} grams.

1 megaparsec equals 3.261 million light years or 9.46 trillion kilometers.

** 1 year equals 31.56 million seconds.

The last stage has a mass of 9,152 electron or 4.98 proton masses.

*** The radius of this compact system is about that of two protons or 6×10^{-13} cm.

In Table I, the theoretical masses are calculated by dividing the theoretical mass of the universe of 7.95×10^{24} solar masses by 2^{10} or 1,024 for each major one of 27 stages. The theory leads to a highly regular pattern in general agreement with the typical total mass values of the major astronomical structures when corrected for the estimated invisible or dark mass. The theory predicted that the average mass of a supercluster would be 7.40×10^{15} solar masses, that of galaxies such as ours, 7.23×10^{12}, that of dwarf galaxies 7.06×10^{9}, and that of the average stellar system 6.58 times the mass of the Sun, which is a star of average mass. These values are about five to ten times larger than the typical visible masses observed, in agreement with the fact that only a small fraction of the total mass in the universe is in visible form.

Given the sizes of the various systems, it is possible to obtain their rotational periods, also shown in Table I. The radii of the space occupied by the rotating electromagnetic field of the initial electron pair extend out to the point where their tangential velocity reaches that of light. As indicated in Figure 14.2, for the present relativistic model this radius is also the Schwarzschild radius of the closed space associated with the mass of the initial pair. The time required for a complete rotation is given by the circumference $2\pi R_s$ divided by the velocity at the outer limit of the system, namely that of light c. Since the radii of the successively smaller systems are smaller than that for the Lemaître atom by the square-root of the mass difference of 1,024, or by a factor of thirty-two for each stage, the periods of rotation rapidly decrease as the mass declines, so that they rotate more and more rapidly.

As shown in Table 1, for the seed of the largest mass, or the primeval atom, the period of rotation is 1.56×10^{13} or 15.6 trillion years. For the seed-masses of the next smaller system, each of which had a mass of 7.76×10^{21} solar masses, the period of rotation was 487 billion years. For the next smaller system with a mass of 7.58×10^{18} or a million trillion Suns, corresponding to 1,024 superclusters, the period of the initial pair was 15.2 billion years, and for superclusters with a mass of 7,400 trillion Suns, it was 476 million years. For galaxies of the typical Milky Way size containing a total mass of 7.23 trillion stars, the seed pair revolved once every 14.9 million years, while for the so-called dwarf galaxies with a total seed-mass of 7.06 billion solar masses, it took only 465,000 years for a complete revolution of 360 degrees.

These rotational velocities are in general agreement with those for the inner regions of spiral galaxies studied by Rubin and Ford, where the velocity rises nearly in a straight line with distance from the center, characteristic of rotation at a constant angular velocity as expected for the case of a rigid field structure such as that of a massive seed pair from which the galaxy evolved. When the velocity reaches the value expected for rotational equilibrium according to Kepler's law for rotation around a central mass, it levels off to a nearly constant value with increasing radius, instead of declining as it would as the square-root of the radius if there were no halo with a large amounts of dark matter. This flattening of the velocity versus radius for spiral galaxies measured by Rubin and Ford represented the evidence for such a halo. The limiting velocity where it stopped changing with radius was a measure of the mass of the galaxy, and I found that these limiting velocities fitted the model.[4]

Using the theoretical value of the mass for the average galaxy of the size of the Milky Way, the expression for the equilibrium velocity gives the predicted average value as 207 kilometers per second. It is 174 kilometers per second for galaxies that have half the average mass, and rises to 246 kilometers per second for a mass two times greater than the theoretical average, in agreement with the typical range of values measured by Rubin and Ford. A factor of two in mass thus leads to a difference of 36 kilometers per second, and a factor of four from the mean to a difference of 72 kilometers per second. This explains the surprising finding by W. G. Tifft in 1976, widely disbelieved at the time, that the rotational velocities of spiral galaxies seem to be "quantized" in steps involving integral multiples of 36 kilometers per second, since the masses of galaxies and all other astronomical objects are not randomly distributed around a mean value, but differ by discrete factors of 2, 4, 8, 16 from the mean value. Thus, the quantization of redshifts discovered by Tifft represents further support for an evolution of the universe in series of divisions by two of a primeval atom. As a result of this quantization of masses in multiples of two, the sizes and thus the spacing of galaxies in a supercluster are also quantized in discrete steps by a factor equal to the square-root of two or 1.41. Since different distances of galaxies from the center of the supercluster mean different radial velocities and therefore different redshifts, this also explains the mysterious quantization of redshifts found by Tifft and Cocke in 1984, and more recently by W. M. Napier and B. N. Guthrie in 1993 for galaxies in our rotating and expanding supercluster, supporting the early findings of Rubin and de Vaucouleurs.

Applying the formula for the equilibrium velocities to the theoretical value for superclusters of 7,400 trillion solar masses, the value of 1,171 kilometers per second is in surprisingly good agreement with the value of 1,000 kilometers per second arrived at by Rubin in her thesis at Cornell, which no one wanted to believe when she first presented her findings at a meeting of the Astronomical Society. But in the last decade, increasing evidence has accumulated for large-scale "streaming" of groups of galaxies, all moving mysteriously together towards what appears to be a "great attractor" as described by Alan Dressler in his 1994 book, *Voyage to a Great Attractor*. When I asked Dressler at the January 1990 meeting of the Astronomical Society in Washington, after he had given a talk on the subject, whether these coherent motions of galaxies could turn out to be rotational as Rubin had concluded, he told me that this was indeed a possibility.

In the case of the next smaller systems, called globular clusters, most clearly seen in a spherical halo around galaxies each with a total of 6.9 million solar masses, the rotational period declines to 14,500 years. Stellar associations, with a total seed-mass of 6,730 solar masses, have a predicted period of just 454 years, while the average stellar system such as ours with a mass for the entire structure, including invisible components like the outlying Kuiper belt of meteors, are predicted by the model to have a period of 14.2 years. This is not far from the average periods of the six inner planets known to the ancients, and the agreement is even closer when Uranus is included. The periods of the planets from the Sun outward as measured in

Earth years: Mercury, 0.24 ; Venus 0.61: Earth 1.00; Mars 1.88, Jupiter 11.86; Saturn 29.46, Uranus, 84.01; Neptune 164.79 and Pluto 247.7. Pluto has the most eccentric orbit and one that is most inclined to that of the other planets, so that R. A. Lyttleton has suggested that Pluto may be an escaped satellite of Neptune and is therefore left out of the calculation of the average period. The average for the innermost six planets is 7.5 years, while the average for the seven innermost planets including Neptune is 18.4 years, nicely bracketing the theoretical value of 14.2 years. Considering that there are no arbitrary or adjustable parameters involved in arriving at these results other than the four fundamental atomic constants, this is a surprisingly good agreement with the theoretically predicted rotational period of the seed pair that apparently gave rise to our Sun and its system of planets.

Moving to the satellite systems of the large planets, the theory gives 0.443 years or 161.8 days. Again, this is close to the observed average for the thirteen satellites of Jupiter and the ten satellites of Saturn. For the case of Jupiter, the average orbiting period is 296 days, while for Saturn it is 67.9 days, giving an overall average of 182.3 days, only twelve percent greater than the theoretical value.

These results for the masses and rotation periods of the planets in the Solar System and the satellites of the large planets strongly support Lemaître's suggestion that the stars are born from massive seeds just like the universe and the galaxies, producing the matter of which they are composed in a miniature version of the Big Bang and ejecting most of the mass in the seed pair to great distances, giving rise to all the gases, dust and other forms of matter in intergalactic and interstellar space. This would apply both to the first stars formed shortly after the Big Bang some twelve to fifteen billion years ago and to stars born in galaxies that formed long after the Big Bang in a delayed ejection process. A fraction of the material ejected at the birth of a star apparently falls back towards the central twenty percent of the pairs that were temporarily trapped. When this mass is large enough, it starts a fusion reaction that produces the energy of the star. When the mass is too small to ignite a nuclear reaction, as in the case of the planets and smaller systems, the planets, asteroids and meteors we know are formed, many of which must have been ejected to very large distances together with still smaller objects, forming the dark haloes around the galaxies and thus part of the dark matter of the universe. The model therefore provides a possible solution to the present problem of stellar and planetary system formation when there is only a uniform gas-cloud in the beginning, and no massive seed mass to cause a rapid collapse into a star. It also implies that all stars are born with planets surrounding them, some of which are at locations that are neither too hot nor too cold for liquid water to exist on their surface, which may allow bacteria, plants, animals and conscious beings to evolve as here on Earth.

The theoretical values for the rotational periods or angular velocities that the model provides can be used to calculate the Hubble Constant for the universe, and these calculations can be tested against observations. The Hubble Constant relates the velocities in the radial direction away from an observer to the distance in the simple form $v_r = H\ r$. In a rigidly rotating system such as the electromagnetic field of a massive electron-positron seed pair, defining the physical space of a given system, exactly the same relation holds between the tangential velocity at any point and the distance from the center: namely, $v_t = \omega\ r$. Gödel investigated the possibility of a rotating closed Einstein universe, and in a corrected version of his theory I. Ozsvath and E. Schücking showed that such a universe would have to rotate like a rigid body. This rotation is similar to the field of the Lemaître atom and all its fragments, each of which is a miniature universe—which ultimately becomes a rotating closed Einstein universe. In the limited case of completed division and expansion, all the various structures end up as miniature rotating Gödel universes. If one now assumes that the simultaneous expansion and rotation result in a homogeneous, expanding system in which shapes, and therefore relative velocity components of objects in different directions, are preserved without distortion, then the radial and tangential velocities must have the same dependence on distance away from the center. If this is indeed the case, then it must be that the angular velocity ω equals the Hubble Constant h.

But the angular velocity ω of the various cosmological seed pairs is given by the speed with which the outermost field-lines move in a tangential direction, namely that of light as shown in Figure 14.2 in Chapter 14, divided by the radius of the system.

In Table I there are only two of the large systems of galaxies whose orbital periods, given by $2\pi / \omega$, are longer than the time since the Big Bang, and must therefore still be expanding. These are complexes of superclusters with a radius 2.42×10^9 light years, and the largest of all systems of galaxies, the supercomplex of radius 77.6×10^9 light years, thirty-two times larger, of which there is only one so that it includes all others. Converting the results to the most commonly-used astronomical units—kilometers per second per megaparsec—I obtained the values 12.6 for the largest system, and 402.3 for the smaller one.[5] This bracketed all the recently measured values from the low of 50, favored by Alan Sandage, to the high of 90 of de Vaucouleurs. Taking the so-called geometric average by multiplying the two values and taking the square-root of the product, I found the value to be 71.2. This was in agreement with the value of 80±17 initially reported by Wendy Freeman and her team of astronomers, who worked with the Hubble telescope afters its successful repair in December 1993. The results were updated in 1997 to 73±6, an even closer agreement. The model therefore explains the average value of the observed Hubble Constant and its wide fluctuations, since the rotational motions of the systems of galaxies add or subtract large velocities to the expansion velocity, depending on which side of a supercluster is examined.

At a constant expansion rate made possible by the centrifugal force countering the pull of gravity, the inverse of the Hubble Constant gives the time since the Big Bang: about 13.4 billion years. This is on the same order as the best estimates for the age of the oldest stars of twelve to fifteen billion years. Thus, the centrifugal force due to rotation of every cosmological system removes the present "age crisis" facing the simple Big Bang model, in which there is no rotation to counter the action of gravity, giving a time elapsed since the Big Bang of only 8.9 billion years.

In the last of the twenty-seven stages in the evolution of the universe from the Lemaître atom, when normal neutrons and protons were formed, the model predicted that only some forty percent of the total mass-energy available went into the formation of nucleons, and sixty percent went into accelerating the newly formed nucleons and mesons formed at that moment to extremely high velocities. Because of the large particle velocities in the explosive formation of nucleons, the universe was heated to nearly ten trillion degrees, producing a very hot Big Bang.

I presented the method by which I arrived at this result at a meeting on the nature of dark matter held at the University of Maryland in October 1994. The temperature arrived at was sufficient for nucleosynthesis, so that this major prediction of the standard Big Bang model is incorporated into the present Lemaître model.

The connection between the microscopic world of nuclear particles and the cosmic scale provided by the electron pair model is illustrated by calculating the lifetime for decay to nuclear particles of the electron pair system at the twenty-seventh stage.[6] The electron pair system's mass at this stage is about that of a J/Psi meson created in a high energy accelerator, and also on the order of one period of rotation, or about 1.1×10^{-20} seconds.[7] Its lifetime should be about that for the decay of the J/Psi. It is, in fact, located exactly between the measured lifetime of the J/Psi meson equal to 1.04×10^{-20} seconds and that of the Upsilon, given by 1.53×10^{-20} seconds. The J/Psi did indeed turn out to be the key to the evolution and structure of the universe, as I had suspected it would on that day in November 1974 when I learned of its discovery from the front page of the *New York Times*.

The time it took for various astronomical objects to divide down to ordinary matter particles from the initial state of a massive electron-positron pair can now be calculated. Based on the lifetimes for the twenty-seventh stage and the J/Psi and Upsilon particles, one can assume that one period of rotation of the initial seed mass is required for the successive stages involving movement of the vortex lines from the rim to the center to achieve scission of the half rings, creating lower mass pairs with increasingly rapid periods of rotation. For the Lemaître atom, it means that it must have been about sixteen trillion years from its formation to the Big Bang. For a typical small planet like the Earth, for which the theoretical period of rotation of the seed pair at the tenth stage is shorter by a factor of 2100, the period of rotation is 5.07 days.

A rotating vortex ring apparently divides itself, in a kind of pinching process, to form a pair of equal and opposite charges at the ends that annihilate each other quickly. Two whole vortex rings, each with half the initial mass, are thereby formed. The new rings in turn undergo a scission or division process, in the course of which two pairs of charges form briefly once again. This process can be visualized to occur as a result of a vibration in the plane of a vortex ring, so that two regions located on opposite sides of the ring move together and overlap. There is no net internal motion at that point, so that the lines of vorticity end in this region, creating two equal and opposite charges. This is what Wheeler and Patton characterized as the "tearing of space to create charges," a process that takes place when the field strength or energy density is great enough, equal to the Planck Density. Because the charges attract each other strongly and there is nothing to prevent them from quickly annihilating each other as there is apparently in the proton, they rapidly merge and two whole vortex rings are produced. The enormous mass of the rings attracts them to each other gravitationally and thus they do not fly part. The two rings immediately repeat the process.

Since the rings of lower mass rotate faster by a factor of the square-root of two, successive steps take place in a time period given by a series terms each less than the previous one by the square root of two. This series equals 3.414 times the time for scission of the first ring or pair of charges if more than a few steps are involved.[8]

At any stage of a division, the time to complete the remaining steps to the moment when protons are formed is closely equal to the rotational period, as listed in Table I, ranging from nearly sixteen trillion years for the Lemaître atom to one hundred million trillionth of a second at the stage when neutrons and protons are formed in the Big Bang or in any of the smaller delayed processes of the same type.

By the time the universe had cooled to about 3,000 degrees, when photons were able to move freely without scattering by free electrons, the average energy of the photons in a so-called black-body spectral distribution was a low 0.25 electron volts, or $4.9 \times 10^{-7}\ m_0c^2$, well in the infrared region of the spectrum. For every nucleon there were 2.8×10^9 photons, a number that has remained constant since the photons escaped, and one that is in agreement with the 1 to 3×10^9 photons per nucleon necessary to explain the abundance of helium, deuterium and lithium, as discussed by Edward Kolb and Michael Turner in *The Early Universe.*

Because of the large mass-energy of the initial electron pair seed that accelerated the newly formed nucleons to high velocities and thus heated the universe, the mass of ejected neutrons and protons that recollapsed under the action of gravity to form the average star like our Sun amounts to only about 2.6 solar masses out of the 6.58 in the original rotating electron pair seed. Of the 2.6 solar masses in particle form, including the seeds for all smaller systems, close to two-thirds or 1.6 must have been ejected to very large distances in the enormous explosion that took place when the nucleons were formed, since the average star has a mass of only one solar mass.

Not all stars have exactly the same mass, nor do all galaxies and other categories of objects. Before ejection of the massive pairs from the cluster began, sometimes the number of steps of division by two was not ten but a few divisions more or less, as a result of statistical fluctuations. For instance, when only nine divisions occurred in the formation of a supercluster, there would be only 512 galaxies, and when eleven occurred, 2,048 galaxies would form, with masses twice or one-quarter of the average value. This agrees with the fact that both stars and superclusters have roughly the same range of masses around the mean, typically from one quarter to eight times the average mass, corresponding to a range from eight to twelve divisions by two around the mean. Thus, the masses of stars and galaxies are distributed around the average value in a pattern that looks like a symmetrical bell-shaped curve when plotted on a logarithmic scale where equal divisions represent factors of two in mass, with very few as low as one sixteenth or as high as sixteen times the average mass. This prediction of the electron pair theory is in fact observed for the distribution of the masses of stars and galaxies, which has such a bell-shaped form when plotted by their absolute magnitude—a logarithmic measure of their luminosity related to their mass.

Continuing the division process beyond the stage where stars were formed, the next or ninth stage gave a mass closely equal to five to twenty times that of the major planets like Saturn and Jupiter, with masses roughly one-thousandth that of the Sun. As an example, Jupiter has a mass equal to 1/1,047 of the Sun, very close to the value 1/1,024 required by the model, or 1.90×10^{30} grams, consistent with the theoretical value of 12.8×10^{30} grams when the ejection of a large fraction of the original mass in the seed pair is taken into account. The following stage resulted in objects with a mass typically a little larger than that of planets such as Venus and Earth, about a million times smaller than that of the Sun, again in agreement with observations. Thus, Venus has a mass of 4.87×10^{27} grams, and the Earth a mass of 5.98×10^{27} grams, while the model predicts a mass of 12.48×10^{27} grams.

It seems that the masses of the planets, their satellites, the asteroids, and the meteors that make up our solar system were all laid down in embryonic form before the time of the Big Bang, and that most of the initial seed mass was shot out at high velocities to great distances, just as it was for the stars. This explosive ejection at every major one of the twenty-seven stages explains why only a handful of planets exist around the Sun, and so few satellites around the planets. Apparently most of the seeds of planets and smaller objects were ejected to great distances, comparable with the distance between galaxies, where the massive gravitational pull of an entire galaxy or small cluster of galaxies prevented escape to still larger distances, accounting for halos of dark matter. The masses of stellar and planet-sized objects were too small to hold back the fragments by gravity when the final stage of division was reached and the explosive formation of nucleons took place. Only the enormous masses of galaxies had gravitational fields strong enough to prevent these high velocity objects from escaping still further.

The ejection of most of the matter in the seeds of solar system mass also explains why the angular momenta of stars and planets are so far below the theoretical expectations based on the pattern for the large systems like galaxies. The non-luminous material ejected at high velocities to millions of light years is simply not visible. Although the ejected mass made up most of the original total mass and also of the angular momentum—proportional to the distance from the center of rotation—the ejected mass was lost in the vastness of space. Thus, the propulsion of many fragments to great distances with masses too small to form luminous stars explains both the abnormally low angular momenta measured for stars and also the large dark halos around all the galaxies.

In addition, entire stellar associations and larger systems such as galaxies that contain by now many burned out white dwarf stars, neutron stars or black holes would also have been ejected from the central clusters in the largest systems. These old systems would today be too faint to be readily seen against the background light of the night sky. This would explain the recent discovery of stars far from any galaxy, as well as a surprising number of very large, dim spiral galaxies containing only very old stars apparently ejected from the centers of superclusters at the time of the Big Bang.

By 1994, I had worked out the details of the embryonic structure of the universe that accounted for yet another problem faced by the standard Big Bang model. In the beginning of this model, there is an infinitely dense flash of light and a mixture of all forms of matter-particles which ultimately includes all the neutrons and protons that now exist. Several minutes later, the temperature dropped to a few million degrees and the formation of the first elements such as helium began. But the detailed theory for this process, known as nucleosynthesis and discussed by Kolb and Turner, accounts for the observed abundance of helium in the universe (about twenty-five percent of all the mass) by stipulating that the number of neutrons and protons, or the number of baryons, could only have been a few percent of the mass needed to achieve a closed universe today. Thus, the theory of nuclear fusion reactions seems to require that most of the matter in the universe cannot be of ordinary baryonic form. This is why so many as-yet-undetected new kinds of particles have been proposed by particle theorists in the last two decades. But despite many searches in high-energy experiments and among cosmic rays, none have been found, accounting in large part for the present state of unease in cosmology and particle theory.

This problem can be resolved by the electron pair model for the origin of matter. As a number of astronomers had been finding for decades, ejections of massive objects from the centers of galaxies seem to be continuing to this day. Observations strongly indicate that at least a fraction of the particles in the original cluster of seed pairs must have remained trapped for many millions or even billions of years beyond the Big Bang to form massive central nuclei capable of giving birth to objects as massive as quasars and dwarf galaxies. This retention of a fraction of the seed pairs can be explained as resulting from the immense masses and small dimensions of the shell into which they were packed. As a result, the slowing-down of all physical processes required by Einstein's theory of general relativity would in effect "freeze" these massive seeds of future galaxies for slow ejection over very long periods of time, explaining why new galaxies beginning as quasars were forming billions of years after the Big Bang. In each cluster of the typical 1024 pairs, some eighty percent are forced out immediately, and the rest remain trapped to form the nuclei of all objects such as galaxies, stellar clusters, stars, planets and so on before the Big Bang. As the 27th stage was reached, only a percent or so of all the matter in the universe was available in the form of nucleons ejected at high energy leading to the great heat needed for the formation of the first elements. The rest were still trapped in the clusters of massive electron pairs in the centers of the systems down to planets and asteroids with rotation rates (and thus lifetimes) greater than a few minutes. As a result, when the initial formation of helium and other low-mass elements such as lithium occurred a few seconds or minutes into the Big Bang, the process of element formation involved only a few percent of all the mass in the universe.

I described these findings in the paper published in the proceedings of the conference on dark matter in October of 1994. The abundance of helium, deuterium, lithium and a few other low-mass elements cannot be explained with the present total amount of matter in the form of ordinary nucleons or in baryonic form but is required by the electron pair theory of matter and the origin of structure in the universe. The amount of matter in the form of ordinary baryons at the time of formation of the elements in the Big Bang was a small fraction of all the mass in the initial fireball, a gigantic hot neutron star roughly the size of our planetary system. Mixed in with the mass of ordinary matter were the portions of the seeds of all the systems of the universe we see today that had not yet been ejected, ranging from gigantic superclusters to galaxies, stars, and planets.

The whole mass of the universe in both protons, neutrons, mesons and the radiation formed by the decay of the mesons into neutrinos and ordinary photons was all expanding together with the seeds of the large astronomical objects embedded in it, like raisins in a rising cake. The growing centrifugal force was due to rotation of the entire field structure of the massive initial pairs at every stage, just as for the original Lemaître atom. And since the seeds of the largest systems containing about a fifth of their original mass in a spherical shell evidently persisted for a very long period, they gave rise not only to the gradually evolving spiral arms of galaxies, beginning at the time of the Big Bang as Ambartsumian had maintained, but also to the ejection of quasars and curvilinear chains of galaxies seen at much later periods closer to our present epoch, as had been argued by Halton Arp and others for decades. Without a physical theory for such processes, these findings were either ignored or regarded as due to coincidental positions of distant objects.

The idea that a fraction of the original massive seed pairs remained trapped in the centers of the largest cosmological structures also explains the peak in the number of quasars found at roughly one-half of the distance back to the Big Bang, together with indications of young galaxies—made up of stars emitting blue (indicating high temperatures) light—at about the same distance. That the largest, most massive systems had the greatest amount of mass for a given amount of light emitted, tends to support the idea that they not only contained large amounts of ordinary matter in a form too faint to be detected, but also that their massive dark nuclei still contain clusters of primordial relativistic electron pairs waiting to start their journey through the cosmos.

These phenomena involving the ejection of matter from the remaining fraction of electron pairs from the centers of the large astronomical structures that exploded with enormous releases of energy were miniature versions of the Big Bang. Matter in the form of ordinary neutrons and protons was still being created from massive electron pairs trapped in an inner space long after the Big Bang, in a way combining aspects of the theory of continuous creation of Gold, Bondi, and Hoyle with the standard Big Bang model going back to Friedman, Lemaître and Gamow.

The Lemaître model also shared an aspect of the model for the early universe before the formation of stable protons, proposed by Guth, in that it has a period of "inflation," but one that took place over a vastly longer period. In the Lemaître model the expansion was driven by the energy contained in the initial electron pair system.This stored-up energy was released by a series of division processes rather than by the emission of radiation, as takes place in normal atoms. Combined with the existence of a minimum approach distance between electrons and positrons, it led to an accelerating expansion or "inflation" that ended only when the first protons and neutrons had been formed. This initial stage was followed by an expansion at a nearly constant velocity as the outward directed centrifugal force countered the inward directed force of gravity. But it did not start from an infinitely dense singularity as the model of Guth, for which the laws of nature break down, but from a state of finite density in the form of a primeval atom composed of a single electron-positron pair, from which all the particles in the universe have descended.

It was the problem of the structure and stability of the protons created in the Big Bang that Luis Alvarez had raised which remained a puzzle. Another question was the structure of the neutron that had been the starting point of the years of effort to understand the nature of matter. These were problems that for a long time seemed to be aggravated by the theory for the successive divisions of the Lemaître atom, namely the increase in the local space curvature with decreasing mass that brought the local value of the gravitational constant to equal the vastly larger electrostatic force. This increase was helpful in understanding the stability of the electron and positron, but if it applied to the proton, as it would have to since it was composed of electrons and positrons in this model, then there would be a serious dilemma.

Since all matter is made of protons, the gravitational attraction between all objects would be far too strong to agree with the extreme weakness of the force of gravity between all macroscopic objects actually observed. Somehow, the huge space curvature associated with the rapid rotation of the electron pair systems in the mesons of which the nucleons were composed had to be neutralized. I came to realize that this would happen if the two sets of pairs held together by a positron exchanged between them were to rotate not in a parallel fashion as I had assumed when I spent a sabbatical leave with Hofstadter at Stanford, but in opposite senses. Not only would this lead to a cancellation of the motions in the fluid ether that produced the local curvature of space, but it would lead to the strongest possible binding of the two sets of electron-positron pairs by the positron exchanged between them, as shown in Figure 15.2 below.

When the two sets of electron pairs on either side of the proton rotate in opposite directions and the positron's motion is carefully synchronized with that of the pair-systems such that its motion is always in exactly opposite direction to the electrons, then there will be the largest relative velocity between them at all times, and thus the strongest relativistic increase in the normal Coulomb Force. This can only happen when the positron moves in a figure-eight orbit, each half of which has to be of the size of the electron-positron orbits. Thus, the relativistic mass of the positron must match very nearly the relativistic mass of the pair-systems, each of which has the mass of a spin 1 neutral pion so as to be always in a position of maximum relative velocity and force of attraction.

At the same time, the magnetic moments of the two pair systems on

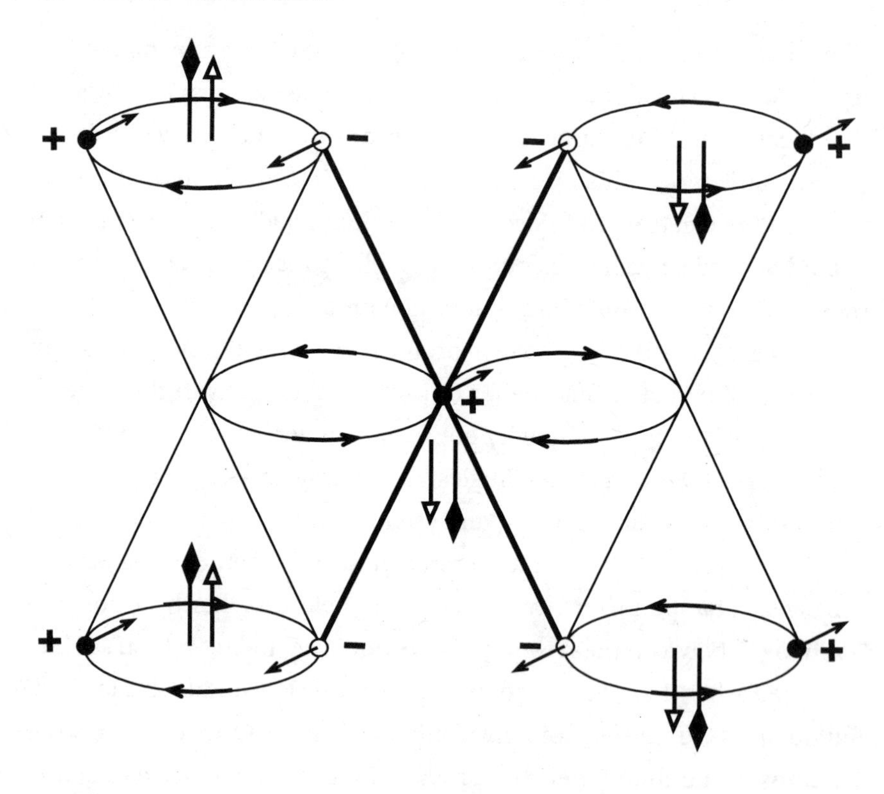

Figure15.2 Schematic diagram of the structure of the proton showing how a positron in a figure eight orbit results in the strongest possible binding of the two sets of electron pairs in the form of K-mesons when they rotate in opposite directions in a synchronized fashion. This can only happen if the orbits all have nearly the same size, so that the positron always moves in the opposite direction to the four electrons, producing the largest possible relative velocity and thus the strongest possible attractive force between these charges. Open arrows show the direction of the net spin of each pair, and the solid arrow the direction of their net magnetic field.

either side of the proton are oriented in opposite directions, which for the parallel orientation required to produce the strongest possible relativistically-enhanced electrostatic force also leads to the strongest magnetic attraction. Thus, after any rotational excitation, the magnetic forces will bring the two sets of pairs back into the state where the binding forces are strongest. Furthermore, the opposing magnetic moments lead to a zero net magnetic moment for the two pairs. Thus, the only spin and magnetic moment that will be measured for the proton will be that due to the positron, giving a net spin $\hbar/2$ to the proton as a whole, in agreement with observation. The strength of the magnetic moment of the positron is reduced from its normally large value when at rest to a very small value, since the size of the positron shrinks due to its enormous relativistic orbital motion, resulting in a value for the proton as a whole of the observed size.

The existence of a relativistically moving positron in the proton thus not only explains the strength of the whole structure, its spin and its magnetic moment together with the spatial distribution of the positive and negative charge found by Hofstadter. But the existence of a positron in the model also explains why the proton has the same magnitude of electrical charge as the electron, something that is not understood in present particle theories. But a proton constituted only of positrons and electrons also achieves the aim of present Grand Unified Theories in that the quarks involved in producing the strong nuclear force are composed of massive, point-like electromagnetic particles with the same spin as the partons or quarks they comprise. A composite nature of quarks has in fact recently been suggested by high energy collisions of protons and anti-protons at the Fermi National Laboratories. And recent experiments at the large electron-proton collider in Hamburg, Germany, have found what appears to be a particle in the proton that acts as if it were a positron with the large mass of a quark—what the electron pair theory requires.

Since the distortion of space (by means of which Einstein explained the action of gravity in his General Theory of Relativity) applies to all forms of matter, the electron pair model automatically realizes the aim of supersymmetric theories seeking to connect gravity with the other forces of nature.

But there is more to this particular structure that makes the proton so stable compared to the J/Psi that also has the same size and square shape, involving two sets of neutral pions held together by a particle in its center. In the case of the proton, this is a single positron that can move extremely rapidly so as to synchronize itself with the rotation of the electrons in the pairs and thus to be as close to them as possible, given the existence of a minimum approach distance between a positron and electron. In the J/Psi, however, there is a heavy pair of particles essentially at rest in the center. As a result, the binding strength of the bonds to the four pairs in the corners of the square structure is much less, and there is no possibility of a highly correlated motion like that in a classical ballet, as in the case of the positron in the proton being constantly exchanged between two partners.

It is this highly correlated motion between all the components of the proton that makes the positron behave in an analogous way to electrons in a superconductor, where electrons can move forever from atom to atom without any electrical resistance. The proton seems to have aspects of such a crystalline structure in which a charge can move indefinitely without losing its energy. It appears to be a perfect example of a system where the properties of a complex structure are not readily deducible from those of its parts, the whole transcending the properties of its constituents. But to understand this, it is necessary to follow Einstein's belief that one must not give up detailed, geometric, space-time descriptions of atomic phenomena.

As for the neutron, it appears to be the kind of structure originally imagined by Ernest Rutherford some seventy years ago: a stable proton to which an electron can attach itself briefly by powerful electromagnetic forces in an orbit of spin $\hbar/2$, being constantly exchanged between neighboring protons in the nucleus of atoms.

The nature of the phase transition that took place when the neutrons and protons were created now also seems clear. From a primordial collection of heavy electron pair systems, a restructuring apparently takes place in which two sets of equal and opposite charges that are strongly bound to each other are paired in such a way that they rotate in opposite directions, held together by a single positron that creates an unbreakable bond between them. Yet it is a bond that provides the freedom for each set of pairs to rotate independently and thus to absorb any amount of energy, first by rotation of the two sets of charges as a whole, and then by the excitation of each electron-positron pair to higher states of energy without any upper limit, apparently providing absolute stability against disintegration. The arrow of time is irreversible. Once formed, the proton can apparently never be destroyed, and the cosmos can evolve only in one direction, forming self-reproducing systems of ever greater complexity, ending with conscious beings trying to understand the mystery of their origin.

Because of its particular architecture, the proton takes on many characteristics of a living system. It is stable in the most diverse environments, yet it has internal degrees of freedom in its motions. It can absorb energy from its environment, and turn this energy into other forms such as massive electron pairs emerging as mesons, returning to its normal state in the process. And when given enough internal excitation energy, it can reproduce itself, giving birth to a proton/anti-proton pair.

Matter is not dead and inert, but a seat of permanent motion. The very first form of matter, a single primeval electron pair, was probably formed from a spinning vortex in an ideal fluid, the underlying substrate or space-time continuum in which all entities are but forms of motion, beginning with a single vortex ring of light.

From this beginning, the universe appears to have evolved much like a living organism, composed of cells that multiply in a series of fission processes, developing discrete structures from systems of galaxies to stars and planets, and ending with stable protons that form the nuclei of atoms able to organize themselves into molecules that allow organic forms and conscious life to come into existence.

Because the proton appears to have a structure that is absolutely stable, matter in its present form can continue to exist forever in an expanding universe. The cosmos is now in its youth, still in the early stages of approaching a finite ultimate size, stabilized by its rotation like all the cosmic structures of which it is composed. And the structure of the proton, as that of the universe as a whole, is determined by just three basic constants of matter that allow mass, space and time to be quantified, together with a single pure number—the fine-structure constant—whose precise value could not differ from its actual value by more than a few percent if life was to come into being.[9]

Since the motional energy which constitutes the cosmos is completely conserved, and since the energy in the movement of galaxies can be converted into new stars being born in the course of galactic collisions, there appears to be no limit to the continuing existence and evolution of life in the universe in all its varied forms.

EPILOGUE

SHORTLY AFTER I SENT OFF the last chapter of this book to the publisher, the July 1997 issue of the journal *Physics Today* arrived with an article describing a recent discovery that greatly supported Lemaître's view of the origin of the universe. Written by one of its editors, Bertram Schwarzschild, it carried the headline "High-Redshift Absorption Lines Show Convincingly that Gamma Ray Bursters Are Very Far Away." Highlighted in a box at the top of the article was the following statement:

> All of a sudden the long debate is over. Gamma-ray bursters really do live halfway across the cosmos. Now we know that they are, for brief moments, the most luminous objects in the universe.

The new data strongly supported my longstanding belief that these energetic phenomena represent the explosive formation of new galaxies and stars from the delayed ejection of massive electron-positron pairs. Such intense bursts of gamma rays were predicted by Lemaître as evidence for his theory that stars and galaxies are born from fragments of a primeval atom rather than from a slow collapse of a diffuse gas. He also stated that the discovery of such gamma rays would confirm the idea that the whole universe expanded from a high density state.

Lemaître postulated that the birth of a star from a primordial seed should be accompanied by the production of a burst of gamma radiation with an energy that represents a significant fraction of the energy associated with the total mass of a star. This energy would be emitted in the first moments of its birth, before the electrons were created in large numbers that would absorb and degrade the gamma rays to less powerful forms of radiation. Now, more than half a century later, it appears that his ideas have been suddenly vindicated by observations in a most dramatic way.

On May 8, 1997, an Italian-Dutch satellite detected a mysterious large burst of gamma rays. This type of emission was first detected in 1969 by a military satellite designed to detect violations of the nuclear test-ban treaty. The latest burst turned out to be the "smoking gun" that would finally decide whether these brief high-energy pulses come to us over cosmological distances or merely from the outskirts of our own galaxy.

Knowledge of the distance from us is absolutely crucial, since the intensity of radiation decreases as the square of the distance. An explosion taking place halfway across the universe has to be a few billion times more powerful than one taking place in the halo of our galaxy involving objects such as meteors colliding with neutron stars.

For the first time, precise identification of the area of the sky where a burst occurred permitted astronomers to detect a visible afterglow. The new ten-meter diameter Keck II telescope in Hawaii was then able to record its spectrum before the phenomenon faded away. Also, for the first time, a record was made of the afterglow that rose to a peak in just two days and then declined to half its maximum intensity in the same time. Using radio telescopes, still another group of astronomers began to see a new radio source at precisely the position of the May 8 gamma burst five days later.

From the large redshifts of the spectral lines it was possible to conclude that the burst must have taken place at a distance on the order of seven billion light years. Knowing the distance, it was calculated that this cataclysmic explosion spewed forth the equivalent of something like the total mass-energy contained in a typical star like our Sun. This is hundreds of times the energy produced by the Sun in its entire life of some five billion years. Yet all this energy was released in the form of gamma rays in a matter of seconds.

This amount of energy has the magnitude of the kinetic or motional energy of the newly created protons in the twenty-seventh stage of division of the Lemaître atom that heated the universe at the time of the Big Bang, as described in the last chapter. This process ultimately produced the cosmic background radiation detected as radio waves by Penzias and Wilson in 1965. The explosion producing the gamma burst is of the same type as that at the moment of the Big Bang, representing the sudden release of the energy contained in the massive seed-pair of a star, instead of that of the whole universe.

From the study of some one thousand such bursts observed in the last twenty-five years, which show spikes of radiation lasting less than a thousandth of a second, it is known that the sources of these bursts cannot be larger than about ten kilometers. This is the size of a neutron star, or a star that has the high density of an atomic nucleus. The duration of the spike in a burst is determined by the difference in the time of arrival of the gamma rays between the radiation emitted at the nearest and the farthest point of the source. This time difference is given by the diameter of the source of the radiation, divided by the speed of light.

The density of the object emitting gamma radiation deduced from the brief duration of a spike in the gamma burst turns out to have exactly the density that Gamow and his students assumed for the universe at the moment of initial formation of the lightest elements in the Big Bang. The universe then consisted of a hot ball of neutrons and seeds of future structures the size of our inner solar system. The objects that are the source of the gamma rays, followed by X-rays, light and radio waves, have all the characteristics of miniature versions of the Big Bang, delayed by a few billion years. It follows that there must indeed be some sort of seeds of galaxies and stars that survived long past the Big Bang to give rise to the birth of galaxies and stars. In the absence of any other such enormously dense forms of matter, it is highly likely that these surviving cosmic seeds were high energy states of the same relativistic electron-positron system that explains the neutral pi-meson, the J/Psi and all other massive particles of matter created in high energy accelerators.

If further studies of the gamma bursts keep supporting the evidence that they occur at large cosmological distances, and that their observed properties keep showing the subsequent emission of X-rays, visible light and radio waves, known to be produced by quasars and active galactic nuclei, then they represent the birth of galaxies and stars that Lemaître had envisioned.

These latest observations in support of a Lemaître-type of Big Bang followed by delayed "mini-Bangs" giving rise to gamma bursters do not stand alone. They are reinforced by recent evidence that there is a peak in the number of newly forming galaxies with blue light emitting young stars as reported in the May 30, 1997 issue of *Science*. This peak, at a distance of about seven billion light years, represents independent support for a delayed production of galaxies long after the Big Bang that cannot be explained by the simple standard model. A few other points in support of the Lemaître model:

- A recent finding by Arp and his colleague Hans-Dieter Radecke at the Max Planck Institute in Garching, Germany, used X-ray emissions to detect two quasars positioned symetrically opposite each other across the centers of 12 out of 24 active galaxies. As summarized in the November 22, 1996 issue of *Science* by Govert Schilling, this finding lends further support to the many previous findings by Arp and others that quasars and small galaxies appear to be ejected from the centers of larger galaxies. But it cannot be explained in terms of conventional cosmology. As Schilling puts it, "If he is right, the implications would be as revolutionary as those of Galileo's claim [about the existence of satellites of Jupiter], which supported the idea that the Earth orbits the Sun just as Jupiter's moons orbit the planet."
- The abnormally high redshifts measured for the symmetrically-located quasars compared with the nearby galaxy are also not understandable within the framework of the standard Big Bang model. However, they can be explained by the high local values of the gravitational constant of the massive electron pairs in the center of a quasar, as required by the Lemaître model and Einstein's General Theory of Relativity. Such gravitational redshifts of quasars near active galaxies always add to the normal redshifts, due to the Hubble expansion velocity. As Schilling says, "Current physical theories cannot explain such an effect," whereas the present Lemaître model requires them.

- The argument for the existence of delayed, small "mini-Bangs" is also supported by the occurrence of a peak in the number of quasars that exists at about the same distance from us as the peak in newly forming galaxies just discovered. Moreover, the Hubble Space Telescope's observations of quasars, reported by the Princeton astrophysicist John Bahcall and others, indicate that they sometimes have almost no detectable galaxies surrounding them. It then becomes difficult to maintain that they arose from the collapse of an existing galaxy to form a central black hole as presently believed. But if they represent an early stage of newly forming galaxies ejected from the center of a larger galaxy as Arp and Radecke's findings indicate, no significant normal galaxy needs to exist around a quasar to explain its formation. Instead, the surrounding galaxy would be expected to arise from a central seed-pair, producing a spherically symmetric mini-Bang associated with the birth of a galaxy, so that only the older quasars would be surrounded by a more fully developed galaxy. The finding that the quasars are also found in closely spaced, abnormally shaped galaxies presently attributed to collisions could now be regarded as due to the ejection of newly evolving pairs and triplets of galaxies from a central source, something that could not be readily accepted by most astronomers in the absence of a physical theory for such a phenomenon.
- Very few spiral galaxies exist near where the peak of new galaxy formation has been observed. According to the Lemaître model, spiral arms are a characteristic of older galaxies, ejected from a central cluster of massive electron pairs. The Hubble Deep Field images taken in December 1995 have shown not only relatively few spiral galaxies where the new galaxies with bright, young, blue stars are forming but also many abnormal shapes rarely seen at closer distances. These galaxies are most often smaller, more nearly spherical and denser than those near us. They frequently occur in the form of chains like pearls on a string, shapes that support recent ejection from a small, dense central source.
- But perhaps the most striking single piece of data suggesting a delayed ejection of galaxies from a rotating central source comes from

the Hubble Deep Field survey. A group of astronomers led by Judith Cohen and using the Keck Telescope reported a remarkable clustering of redshifts or distances of galaxies in the *Astrophysical Journal* in November 1996. The clustering at certain distances resembles that found by Broadhurst in a deep, "pencil-like" survey discussed earlier, and was very recently supported by a large-scale study in a paper published by J. Einastos and an international group of astronomers in the January 9, 1997 issue of *Nature* which showed a regular grid-like spacing of superclusters.

In the largest peak of Cohen's study at a redshift of 0.475, corresponding to a distance of about seven billion light-years, some 18 galaxies were found to lie in a plane at right angle to the line of vision. As shown below, these young galaxies are arranged in a striking curvilinear pattern at regular intervals, consisting of two S-shaped spiral arms, just as one would expect if they were ejected from a closely spaced pair of rotating central sources.

This arrangement of galaxies in two almost perfectly flat parallel planes could not have been produced in the same type of explosion as took place in the Big Bang, which resulted in a spherically symmetric or isotropic form of the universe, and led to the nearly spherical form of the newly ejected galaxies themselves. It could only result from a gradual, periodic escape of objects from two rotating central sources in the center of an evolving supercluster. Rotating shells of massive, primordial electron pairs, each containing the seeds of about two hundred galaxies remaining from an initial number of a thousand or so, gradually collapsing under the action of gravity and ejecting the seeds of galaxies one by one could explain both the extraordinarily flatness of the planes in which the new galaxies appeared and their highly regular spacing.

The luminosity of these galaxies, most frequently forming sets of two or three like double or triple stars, increases regularly with distance from the center. This is just what one would expect for newly developing and expanding young galaxies ejected from a massive dark central cluster. None are fully evolved normal spiral galaxies. But they have a nearly

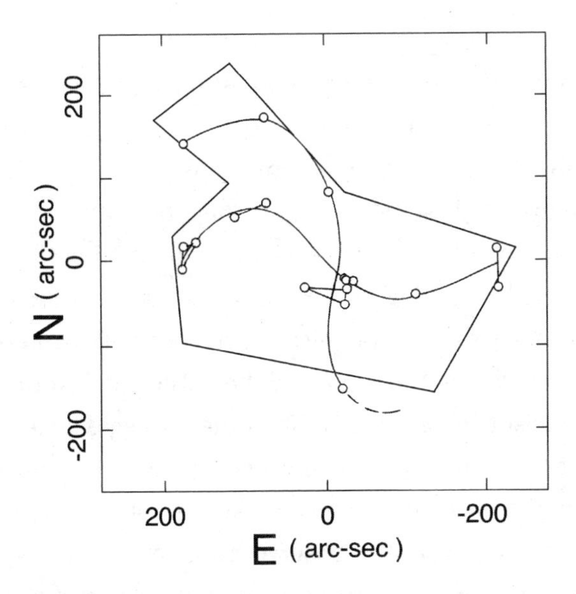

Spiral-like pattern of newly evolving galaxies as reported by Cohen and co-workers in the Hubble Deep Field at redshift 0. 475 or some 7 billion light-years. The centers of all the groups of galaxies in this pattern lie within 2% of the mean redshift or distance of the S-shaped lines to which they belong, and within only 0.4% in the case of the east-west structure. The plane of the vertical or north-south spiral is located behind the east-west, or horizontal one, by 0.5% of the distance from our position.

spherical or anomalous shape, exactly what an origin of the galaxies from a central rotating fragment of the original primeval atom requires.

The hypothesis that all cosmological systems rotate, including the Lemaître atom or the universe, has also been supported by an article published in the April 21, 1997 issue of *Physical Review Letters.* John Ralston and Borge Nodland used a different type of evidence based on observations in the radio spectrum. The article suggests that the universe has "a preferred axis," as if it is rotating. This conclusion is based on the observed rotation of the plane of polarization of radio waves traveling across vast distances of intergalactic space, emitted by certain galaxies.

According to Norland and Ralston, the radio waves need to travel about a billion years at the speed of light for the plane of polarization to twist one complete turn. When I looked at the rotation rates of the superclusters (as listed in Table I of the last chapter), I found that they rotate at a rate of a complete turn approximately every half a billion years, while superclusters complete a revolution in about fifteen billion years. Thus, the two rotation rates bracket the observed ones for the systems traversed, the geometric mean being 2.7 billion years. Considering the different orientations of the axes of these systems and other observational uncertainties, the fact that the order of magnitude of the rotation of the planes of polarization and the cosmological systems they traverse is the same further supports a universe as a hierarchical organization of rotating systems.

Although many astronomers greeted the Ralston and Norland paper with great skepticism, when Vera Rubin at the Carnegie Institution was asked about the possibility that the universe might be rotating by Clayborn Ray of the *New York Times* a few weeks later, she answered, "Most astronomers would say no, but an observer's answer is that we do not yet have enough observations to answer the question." She added that there was no known mechanism that would give the universe so much angular momentum, or spin, at its origin and few mechanisms for adding spin.

Rubin was right in saying that there is no known mechanism for producing the enormous angular momenta of planets, stars, galaxies and the universe as a whole, starting with a uniform cloud of gas as the initial state, according to the currently accepted paradigm. But if physical space and time began with a rotating primeval pair containing an enormous energy, which divided by two in a series of steps, with the field of each new pair spinning and expanding like a rigid body, then the angular momenta would increase from their small initial values to the large values observed.

The architect of this design and the energy required to bring this about remain a source of mystery, awe, and wonder beyond the ken of science. Einstein asked whether God had any choice in designing the universe. Only a very small range of values—for the energy of the first quantum, the strength of its interaction, its limiting velocity and its initial angular momentum—make it possible for conscious beings to evolve with the capacity to admire the simplicity and beauty of a design that gave birth to the cosmos as a spinning ring of light and allows it to exist forever.

Chapter 8

1. Davisson and Cunsman were studying the scattering of a beam of electrons from metals, using an arrangement very much like a modern television tube, where electrons are emitted from a hot tungsten filament by a few thousand volts.

2. Heisenberg showed that actual observations with a microscope using photons of wave length equal to the accuracy wiith which the position of a particle is to be determined would impart a certain momentum to the particle. The shorter the wavelength used, or the more accurately the position is to be measured, the larger is the momentum given to the particle. This limits the precision with which the initial conditions needed to predict the future positions of the particle can be calculated. The determinism of classical physics as envisioned by Laplace would thus be rendered impossible. Because of de Broglie's finding that electrons or any other particles would obey the same relation between their momentum and wavelength as photons, using an electron microscope in which electrons are used instead of photons would not change this conclusion.

3. Bohm and Weinstein concluded that such a classical model, when quantized according to Bohr and de Broglie for the electron in hydrogen orbits, could lead to the existence of more energetic states whose masses were integral multiples of π (1/a) m_0c^2 or n x 430 m_0c^2. The symbol a represents the so-called fine-structure constant, a pure, dimensionless number given by $hc/2\pi e^2 = 1/137.036$ m_0 is the mass of the electron, *e* its charge and *c* the velocity of light. This value of the energy or mass of the lowest excited state, that is its value for the case of n =1, is in fact of the same order of magnitude as the observed mass of the pi-mesons, namely about 274 m_0, or about (2/ α) m_0.

4. The inverse square Coulomb law e^2/r^2 applies, except that the square of the electronic charge e^2 is replaced by γe^2, where γ is the so-called "Lorentz contraction factor," a measure of the degree by which moving objects shrink in the direction of motion, rest-masses increase and clocks slow down according to relativity theory. The result is that motion relative to the ether cannot be detected. The valve of the Lorentz contraction factor is given by $\gamma = 1/(1- v^2/c^2)^{1/2}$, where v is the velocity of a particle relative to the observer, *c* is the velocity of light, and the exponent 1/2 signifies the square root of the relation in parentheses. As v approaches *c*, $1- v^2/c^2$ approaches zero, and γ approaches infinity. The Coulomb Force can thus become very large as the relative velocity increases, balancing the centrifugal force $\gamma m_0 v^2/r$ at all velocities. In effect, in the relativistic relation for the equality of the Coulomb Force—which holds the two particles together while the centrifugal force pushes them apart—the Lorentz factor γ is cancelled, leading to the classical Newtonian relation $m_0 v^2/r= e^2/r^2$. Solving for the radius gives $r= e^2/m_0 v^2$. If v goes to infinity, *r* goes to zero and a singularity results. But if v cannot exceed *c*, then there is a finite minimum radius $r_{min}= e^2/m_0 c^2$ for the electron around a proton.

Chapter 10

1. There thus arises a case where orbital angular momentum, normally quantized in integral units of $\hbar$, is quantized in units of angular momentum $\hbar/2$ for an observer in the precessing frame normally associated with the spin of an electron, positron, proton, or neutrino.

Chapter 12

1. By making the spin opposed to one of the two particles of the central pair-systems, the net spin of the muon could be $1/2\,\hbar$, just as is observed. Calculating the strength of the binding and the relativistic increase in mass of the extra electron finally gave a net mass for the muon of 206.7 electron masses, remarkably close to the observed value of 206.67.

Next, I calculated the life of the muon, which decayed to give rise to an electron or positron, together with two neutrinos. In this calculation, the two radiation quanta were produced by the annihilation of the central charge-pair, exactly as in the case of the spin 0 pion as described in my 1961 paper, except that two neutrinos of spin 1/2 instead of two ordinary photons or gamma rays of spin 1 were formed. I used well-established methods for calculating the lifetime of electromagnetic radiation used in atomic and nuclear physics, the only difference being that the electrons or positrons were now both heavier and smaller due to the highly relativistic motion, vastly lowering the rate of radiation. When I finished my calculations, the value of 2.21 millionth of a second was again in close agreement with the observed value of 2.212.

2. Because the orbiting charge moves opposite to the central pi-meson, the charged pion has half a unit of spin less than the muon, or zero spin. Calculating the corresponding higher energy, once again the agreement with the observed value was very good, 273.5 electron masses versus 273.2 observed. Calculating the life, the theoretical value turned out to be 2.49×10^{-8} seconds, or about two one hundred millionths of a second, as compared with the best observed value of 2.55×10^{-8} second.

3. The so-called Sommerfeld fine-structure constant is a pure number given by the ratio of the square of the electron's charge to the product of Planck's Constant $\hbar$ and the velocity of light c and symbolized by the Greek letter alpha, so that $1/\alpha = \hbar c/e^2 = 137$, or more precisely 137.036.

Chapter 13

1. At this energy, the $n = 3 \times 22$ or the 66th level of the spin 1 neutral pion and the $n = 3 \times 23$ or the 69th level of the spin 0 pion, both located at 18,216 m_0 electron masses. This is again close to the observed mass of three J/Psi mesons, namely 18,180 m_0 electron masses, a difference of only 36 m_0 electron masses, or just 0.2 percent. No such long-lived state is seen at twice the J/Psi mass, even though at this mass energy, the spin 1 and spin 0 energy levels again coincide. The reason is that there is no system of oppositely rotating electron pairs with an excited rotational state of the kind existing for the J/Psi or Upsilon at this energy. The absence of such a state at twice the J/Psi mass is therefore further evidence in support of the models for the J/Psi and the Upsilon, and thus also for the Lemaître atom as a massive electron-positron pair in orbit around each other.

Chapter 15

1. For every neutral pion of theoretical mass $(2/\alpha)$ -10 or 264.07208 m_o there are two spin 1 pions of mass (2/a)+2 or 276.07208 m_o formed, resulting in a weighted average of $(2/\alpha)$-2 or 272.07208 m_o. When I found that when I calculated how many times the mass of the universe obtained from the observed values of *e*, m_o and *G* in the relation $M_u = e^2/m_o^3G^2 = 1.736 \times 10^{85}$ m_o would have to be divided by two to produce two pions of average mass 272.07208 m_o, the answer on the calculator display was 274.070986, almost exactly equal to $2/\alpha = 274.07208$, within the experimental uncertainty of ± 0.00022. It meant that the mass of the universe could be expressed simply as $2[(2/\alpha) - 2]\, 2^{2/\alpha}$ m_o.

2. Solving the Dirac relation $(e^2/m_o^2G)^2 = (M_u/m_o)$ for *G* gives $G = (e^2/m_o^2)(m_o/M_u)^{1/2}$, where the 1/ 2 power represents the square-root. Using the accepted value for $1/\alpha$ listed in the *Particle Properties Data Booklet* of April 1980 equal to 137.03604±0.00011, I obtained for M_u/m_o the value $2[(2/\alpha) - 2]\, 2^{2/\alpha} = 1.736325032 \times 10^{85}$. The square-root was $4.166923364 \times 10^{42}$ and putting in the values for the listed charge of 4.803242×10^{-10} electrostatic units and the mass of $9.1099534 \times 10^{-28}$ grams for the electron, the calculator showed $G = 6.67207785 \times 10^{-8}$, or rounded off to five significant figures, 6.6721×10^{-8} compared with the best observed value listed of 6.6720± 0.0041 in centimeter-gram-second or c.g.s. units, clearly an incredibly close agreement between the theoretical value and the best available measurement at the time.

Note that for M_u equal to m_o, $G = e^2/m_o^2$ so that the gravitational force m_o^2G/r^2 between an electron and a positron becomes equal to the electrostatic force e^2/r^2.

3. This is the Planck length given by $(\hbar G/c^3)^{1/2}$, whose magnitude is 1.616×10^{-33} centimeters when the Newtonian value of *G* is used, the only unit of length that could be formed with $\hbar$, *G*, and *c*.

4. The value for the theoretical equilibrium velocity is given by $(c/2^{1/2})(M/M_u)^{1/4}$ because in the electron pair model the radii of all systems vary as the square root of the mass *M*, the mass of the universe being M_u. This relation between the equilibrium rotational velocity and the mass of a galaxy is similar to the empirical Tully-Fisher relation on the basis of which absolute masses of distant galaxies can be obtained, thus explaining the physical origin of this connection between mass and rotational velocity that can be measured by the Doppler shifts in the spectral lines from the opposite sides of a rotating galaxy whose axis of rotation is close to being at right angles to the line of sight.

5. Since the angular velocity $\omega = c/R_s$, where R_s is the Schwarzschild radius of a system and *c* the tangential velocity at R_s, and the Hubble Constant *H* has the same form as ω, its value for each system should be $H=c/R_s$ if the theory of rotating and expanding astronomical systems is correct. The conversion factor from *H* in kilometers per second per megaparsecs for angular velocity given in radians per seconds is 1 km./sec/Mpc $= 3.24078 \times 10^{-20}$ sec^{-1}, where a radian is the angle 360 degrees/2π=57.3 degrees.

6. As Table I shows, the mass of a relativistic electron pair system at the 27th stage, or the 270th division by two, is the mass of the Lemaître atom or the universe of 1.736×10^{85} electron masses m_o divided by $2^{270} = 1.897 \times 10^{81}$, or 9151 m_o. With a neutron mass of 1838.7 m_o, it takes twice this, or 3677 m_o to form two neutrons flying apart in opposite directions, needed to conserve linear momentum. This leaves 5774 m_oc^2 available for motional energy of the two nucleons and the formation of a few mesons that ultimately decay into fast moving electrons,

positrons, gamma rays and neutrinos. All the energy not needed to form the nucleons, or some 60% of the initial mass-energy of the pair, therefore go into heating the universe, while 40% ends up in the form of stable baryonic matter. Assuming that about half the motional energy and radiation ultimately ends up as photons, the result is that some one quarter of 5774 m_0c^2, or 1444 m_0c^2 of the energy per nucleon appears as heat, which translated into the equivalent of 8×10^{12} degrees centigrade above absolute zero. This is well above that needed for nucleosynthesis, so that the calculations for the production of the lightest elements according to the standard model are seen to apply after the formation of nucleons.

7. The J/Psi that has a mass of 6,060 m_0 compared with the 9,151 m_0 of the state corresponding to the 270th division by two of the primeval atom, and that of the Upsilon at the larger mass of 18,513 m_0. Dividing the period of the Lemaître atom or that of the universe of 1.56×10^{13} years, equal to 4.92×10^{20} seconds by the square-root of 2^{270}, that is by 2^{135}, one obtains 1.13×10^{-20} seconds.

8. The shortest time for division of a vortex ring takes a time equal to the radius of the ring R_s divided by the speed of light with which the portions of the ring approach each other, and it takes an equal time for the rings to regain their normal circular shape. Therefore a good estimate for the time required to produce a single division is $2 \times R_s / c$. The time for a series of successive divisions is obtained by adding up their time periods as they decrease by a factor equal to the inverse square-root of two, or by 0.7071 at every step. This leads to a series of terms whose limiting value for many steps is given by $1 / [1 - 2^{-1/2}] = 3.414$. Thus, even as many as 270 steps take only $6.828\ R_s / c$, which is close to one period of revolution of the initial ring, given by $2\pi\ R_s / c$, since $2\pi = 6.28$.

9. The mystery of the actual value of the fine-structure constant is brought out even more strongly if one examines the precise value of the pure number that governs the nature of the universe which has allowed life as we know it to evolve. It now seems that this would have been possible with values for its inverse anywhere between about 136 and 138. The mystery is why it should have precisely the value 137.036, or, as the most recent precision measurements indicate, most likely the value 137.0360 (to within about one part in a million). The simple integral value 137, or any arbitrary "irrational" number beginning with 137 having an infinite number of decimal places, would have allowed stars, planets, stable protons and the essential elements of carbon and oxygen to come into being. Thus, the appearance of 360 as part of the fine-structure constant, the number of degrees in a complete circle introduced by the ancient Babylonians, is surprising. That number also represents very closely the number of spin revolutions per orbital revolution, or the number of days in a year of a planet such as ours — able to have water on its surface at a temperature that allows a variety of complex lifeforms to evolve. (It is also the length of the sacred year of the Egyptian and the Mayan civilizations.) This value for $1/\alpha$ leads to the result that the mass of the universe in units of the electron mass is given by $2[(2/\alpha) - 2]\ 2^{2/\alpha}$, which is given by $2[(2 \times 137.0360) - 2]\ 2^{2 \times 137.0360}$. This equals 1.73623×10^{85} to six significant numbers, which can also be written as $17.3623 \times 10^{2 \times 6 \times 7}$.

Using the value for $\alpha = 137.036$, the constant of gravitation G, given by $(e^2 / m_0^2) / [2(2/\alpha) - 2]^{1/2}$, becomes $6.672\ 38 \times 10^{-8}$ c.g.s. units, compared with the (1986) best measured value of $6.672\ 59 \pm 0.000\ 85 \times 10^{-8}$ using the 1986 values of the electron mass and charge, $9.109\ 389\ 7 \times 10^{-28}$ gram and $4.803\ 206 \times 10^{-10}$ electrostatic units respectively. The difference in the values of

$1/\alpha$ is only 0.000 21 or 40 times less than the present experimental uncertainty. Thus, the hypothesis that $1/\alpha$ has exactly the rational value 137.036 is in excellent agreement with the best presently measured value of *G*. It will be further tested in the future by improved measurement of Newton's constant to an accuracy of ± 0.00002 or to 3 parts per million, together with the theoretical prediction of the mass of the universe on which the calculation of the gravitational constant depends.

BIBLIOGRAPHY

Asimov, I. *Asimov's Biographical Encyclopedia of Science and Technology.* Garden City, N. Y.: Doubleday, 1964.

Barrow, J. D. *The Origin of the Universe.* New York: Basic Books, 1994.

Barrow, J. D., and F. J. Tipler. *The Anthropic Cosmological Principle.* Oxford: Oxford University Press, 1986.

Bohm, D. *Causality and Chance in Modern Physics.* Philadelphia, PA: University of Pennsylvania Press, 1980.

Bohr, N. *Atomic Physics and Human Knowledge.* New York: John Wiley and Sons Inc., 1958.

Cornell, J. (ed.) *Bubbles, Voids, and Bumps in Time: The New Cosmology.* Cambridge: Cambridge University Press, 1995.

Davies, P. *Superforce: The Search for a Grand Unified Theory of Nature.* New York: Simon and Schuster, 1984.

Davies, P., and J. Brown, eds. *Superstrings: A Theory of Everything* Cambridge: Cambridge University Press, 1988.

De Broglie, L. *Matter and Light.* New York: Dover Publications, 1939

Dressler, A. *Voyage to the Great Attractor.* New York: Knopf, 1971.

Dyson, F. *Disturbing the Universe.* New York: Harper & Row, 1979.

Dyson, F. *Infinite in All Directions.* New York: Harper & Row, 1988.

Einstein, A. *Sidelights on Relativity.* New York: Dover Publications, 1983 (orig. pub. 1922).

Ferris, T. *The Whole Shebang: A State-of-the-Universe Report.* New York: Simon and Schuster, 1997.

Feuer, L. *Einstein and the Generations of Science.* New York: Knopf, 1947.

Gamow, G. *Mr. Tompkins in Wonderland.* Cambridge: Cambridge University Press, 1939.

Gamow, G. *One, Two, Three...Infinity.* New York: Viking Press, 1947

Gamow, G. *The Birth and Death of Our Sun.* New York: Viking, 1952.

Gamow, G. *Thirty Years That Shook Physics: The Story of Quantum Theory.* Garden City, NY: Anchor Books-Doubleday & Co., 1966.

Guth, A. H. *The Inflationary Universe*. Reading, MA: Addison-Wesley, 1997.

Harrison, E. R. *Cosmology: The Science of the Universe*. Cambridge: Cambridge University Press, 1981.

Hawking, S. W. *A Brief History of Time*. New York: Bantam, 1988.

Heisenberg, W. *Across the Frontiers*. New York: Harper and Row, 1974.

Heisenberg, W. *Physics and Beyond*. New York: Harper Torchbooks, 1971.

Kaku, M. *Hyperspace: A Scientific Odyssey Through Parallel Universes, Time Warps, and the 10th Dimension*. Oxford: Oxford University Press, 1994.

Kaku, M. *Introduction to Superstrings*. New York: Springer-Verlag, 1988.

Kant, I. *Universal Natural History and Theory of the Heavens*. Ann Arbor: University of Michigan Press, 1969 (orig. pub. 1755).

Kolb, E. W. and M. S. Turner. *The Early Universe*. New York: Addison-Wesley, 1990.

Kolb, R. *Blind Watchers in the Sky: The People and Ideas that Shaped Our View of the Universe*. New York: Addison-Wesley, 1996 .

Kuhn, T. *The Structure of Scientific Revolutions*. Chicago: University of Chicago Press, 1971.

Lemaître, G. *The Primeval Atom*. New York: Van Nostrand, 1950.

Lorentz, H. A. *The Theory of Electrons*. New York:Dover Publication, 1951 (orig. pub. 1915).

Lorentz, H. A., A. Einstein, A. Minkowski, and H. Weyl. *The Principle of Relativity: A Collection of Original Memoirs of Relativity*. New York: Dover Publications, 1952 (orig. pub. 1923).

Margenau, M. *The Nature of Physical Reality*. New York: McGraw-Hill, 1950.

Pais, A. *Inward Bound: Of Matter and Forces in the Physical World*. Oxford: Clarendon Press, 1986.

Pais, A. *Subtle is the Lord: The Science and the Life of Albert Einstein*. Oxford: Oxford University Press, 1982.

Peebles, P. J. E. *Principles of Physical Cosmology*. Princeton: Princeton University Press, 1993.

Penrose, R. *The Emperor's New Mind*. Oxford: Oxford University Press, 1989.

Popper, K. R. *Quantum Theory and the Schism in Physics*. Totowa, New Jersey: Rowman and Littlefield, 1982.

Poincaré, H. *The Foundations of Science*. Lancaster, PA: The Science Press, 1946.

Sagan, C. *Cosmos*. New York: Random House, 1980.

Scott, J. F. *The Scientific Work of Rene Descartes*. London: Taylor and Francis, 1956.

Shu, F. H. *The Physical Universe: An Introduction to Astronomy*. Berkeley, CA: Interstellar Media, 1987.

Smoot, G., and K. Davidson. *Wrinkles in Time*. New York: William Morrow & Co., 1993.

Trefil, J. S. *From Atoms to Quarks*. New York: Scribner, 1980.

Trefil, J. S. *The Moment of Creation*. New York: Macmillan, 1983.

Weinberg, S. *The First Three Minutes: A Modern View of the Origin of the Universe*. New York: Basic Books, 1988.

Westfall, R. S. *Never at Rest: A Biography of Isaac Newton.*, Cambridge: Cambridge University Press, 1980.

Wheeler, J. A., and C. M. Patton. In E. J. Isham, R. Penrose, and D. W. Sciama, eds., *Quantum Gravity*. Oxford: Oxford University Press, 1975.

Whittaker, E. *A History of the Theories of Aether and Electricity*. London: Thomas Nelson and Sons, 1953.

Whyte, L. L. *The Atomic Problem: A Challenge to Physicists and Mathematicians.* London: George Allen and Unwin, 1961.

Williams, L. P. *The Origins of Field Theory*. Lanham, MD: University Press of America, 1980.

Young, L. B. *The Unfinished Universe*. Oxford: Oxford University Press, 1986.

Zee, A. *Fearful Symmetry*. New York: Macmillan, 1986.

E. J. STERNGLASS:

Publications on Particle Physics and Cosmology

"The Threat of Scientific Obscurity," *The Cornell Review*, Spring 1952.

"On the Evidence for an Extended Source Model of the Electron," *Comptes Rendus*, Paris *246*, 1958, pp. 1166–1168.

"On the Theoretical Problem of Extended Source Models for the Electron and Proton," *Comptes Rendus*, Paris *246*, 1958, pp. 1386–1389.

"Relativistic Electron Pair Systems and the Structure of Neutral Mesons," *Physical Review 122* , 1961, pp. 391–398.

"Particle Interference and the Causal Space-Time Description of Atomic Phenomena," *Horizons of a Philosopher: Essays in Honor of David Baumgardt*, edited by J. Frank, H. Minkowski and E.J. Sternglass, Leiden, Netherlands: E.J. Brill, 1963.

"The Elementary Particle Concept in Modern Physical Theory," *Proceedings Tenth International Congress of the History of Science.* Paris: Hermann, 1964, pp. 355–358.

"Evidence for a Molecular Structure of Heavy Mesons," *Proceedings International Conference on Nucleon Structure.* Stanford University Press, 1964, pp. 340–343.

"Electron-Positron Structure of the Charged Mesons and Meson Resonances," *Nuovo Cimento, 35,* January, 1965, pp. 227–260.

"New Evidence for a Molecular Structure of Meson and Baryon Resonance States," *Proceedings Second Conference on Resonant Particles,* Ohio University, 1965, pp. 33–56.

"Evidence for Relativistic Electron Pair Model of Nuclear Particles," *International Journal of Theoretical Physics, 17,* 1978, pp. 347–352.

A Model for the Early Universe and the Connection between Gravitation and the Quantum Nature of Matter, *Nuovo Cimento Letters , 41,* 1984, pp. 203–208.

"The Quantum Condition and the Non-Euclidean Nature of Space-Time," *Annals of the New York Academy of Sciences, 480,* 1986, pp. 614–617.

"The Origin of Black Holes in Active Galactic Nuclei," in *Testing the AGN Paradigm,* Holt, S. S., S. G. Neff and C. M. Urry, eds. *AIP Conference Proceedings 254,* New York: American Institute of Physics, 1992, pp. 105–108.

"The Relativistic electron pair Theory of Matter and its Implications for Cosmology," *Proceedings of the International Conference Frontiers of Fundamental Physics,* Olympia, Greece, September 27–30, 1993, London: Plenum Publishing Company, 1994, pp. 59–65.

"Evidence for a Possible Common Origin of Baryonic and Non-Baryonic Dark Matter." *Proceedings of the Conference on Dark Matter.* Holt, S. S. and C. L. Bennett, eds. University of Maryland, October 1994, AIP Conference Proceedings *336* , American Institute of Physics, 1995, pp. 513–516.

"Implications of Extended Models of the Electron for Particle Theory and Cosmology," *Proceedings of the Conference on The Present Status of the Quantum Theory of Light* . Toronto, Canada, August 27–30, Dordrecht, Netherlands: Cluver Publishing Co., 1996, pp. 459–469.

Available on www.Pitt.edu/~sterngla

GLOSSARY

Absorption lines. Dark lines in spectra. Produced when light from a distant source passes through a gas cloud closer to the observer.

Alpha particle. The nucleus of a helium atom: two protons and two neutrons.

Angular momentum. Product of mass, radius, and velocity for a rotating object.

Baryons. A class of strongly interacting particles, including neutrons, protons, and the unstable hadrons known as hyperons that decay to protons.

Beta particle. Electron emitted from the nucleus of an atom during radioactive decay.

Black-body radiation. Energy in the form of photons re-emitted by an object capable of absorbing all energy that strikes it.

Black hole. Object with a gravitational field so intense that no light or moving particle can escape from it. In the electron pair theory, there is no singularity at the center.

Boson. Subatomic particle with one whole unit of spin $h/2\pi$.

Causation, causality. Doctrine that every new situation must have resulted from a previous state.

Cepheid variable. Pulsating variable star whose period of brightness variation is directly related to its absolute luminosity. Used to measure distances of galaxies.

Charm. Fourth type of quarks.

Classical physics. Physics prior to the discovery of the particle nature of light and the wave-like properties of the electron. Incorporates Newtonian mechanics and assumes a causal relation for physical phenomena. It also assumes that mechanical, geometric models can be used on every scale.

Closed universe. Cosmological model in which the universe is spherical and has a finite radius from which no light can escape, as in the case of a black hole.

Cosmic microwave background (CMB). Also known as cosmic background radiation. Microwave radio emission consisting of photons released when the universe was about 300,000 years old and hot; now cooled owing to expansion of the universe.

Cosmic rays. High energy charged particles or energetic photons which enter

Earth's atmosphere from outer space. Some originate from distant regions, some from our Sun.

Cosmological constant. A term added by Einstein in 1917 to his gravitational field equations. Such a term repels and would be needed in a static universe to balance the attraction due to gravitation. Explained by centrifugal force in the present electron pair model, where every cosmological system rotates.

Dark matter. Matter whose existence is inferred on the basis of dynamical studies — for instance, the velocity of stars and gases around galaxies — but which does not show up as bright objects. It may be of ordinary or baryonic type such as small planets, ejected to great distances, or it can be composed of relativistic electrons and positrons.

Density. The amount of any quantity per unit volume. The mass density is the mass per unit volume.

Determinism. Doctrine that all events are in principle predictable effects of prior events, given perfectly defined initial velocities and positions of all particles.

Deuterium. A heavy form of hydrogen, H^2. The nuclei of deuterium, called deuterons, consist of one proton and one neutron.

Dirac equation. Mathematical description of the electron that incorporates both quantum mechanics and special relativity.

Doppler Effect. The change in frequency or wave length of any signal, caused by a relative motion of source and receiver.

Dwarf galaxy. A system containing about a million stars.

Electromagnetic force (or interaction). Fundamental force of nature that involves electrically charged particles.

Electron. The lightest massive elementary particle. The chemical properties of atoms and molecules are determined by the number of electrons in atoms.

Electron volt. A unit of energy, convenient in atomic physics, equal to the energy acquired by one electron in passing through a voltage difference of one volt.

Ether. A fluid-like medium pervading space.

Expansion of the universe. Increasing distance between galaxies at a rate that grows with the galaxies' distance from us or observers in any other galaxy.

Field. A region of space in which the action of a force on a particle can be described mathematically.

Fine-structure constant. Fundamental numerical constant of atomic physics. Defined as the square of the charge of the electron divided by the product of Planck's Constant, 2π, and the speed of light. Denoted by α, equal to 1/137.036.

Fission. A process in which heavy atomic nuclei split into two or three pieces, releasing energy.

Flavor. Designation of quark types — up, down, strange, charmed, top, and bottom.

Frequency. The rate at which crests of waves pass a given point, equal to the speed of the wave divided by the wavelength. In general, the number of oscillations per second of objects. It is measured in cycles or number of vibrations per second.

Galaxy. A large gravitationally-bound cluster of stars, containing about a trillion or 10^{12} solar masses. Galaxies are generally classified as elliptical, spiral, or irregular. All galaxies rotate and evolve. In the Lemaître model, they begin as massive electron-positron pairs that give rise to quasars and then elliptical galaxies which gradually develop spiral arms.

Gamma rays. Photons of very large energy, greater than X-rays.

General Relativity. The theory of gravitation developed by Albert Einstein in the decade 1906-1916. The essential idea of the General Theory of Relativity is that gravitation is an effect of the distortion of the space-time continuum, or ether. It also assumes that even for observers in accelerated reference frames, all the normal laws of physics apply.

Gluons. Massive photon-like entities that have integral spin and carry the strong nuclear force. In the electron pair theory, they are vortex rings of great energy like photons, but they cannot escape from protons, where they are produced in high energy collisions.

Gravity. Attractive force experienced by all particles, including photons, caused by the distortion of space.

Hadrons. A class of elementary particles, divided into baryons and mesons. Hadrons respond to the strong nuclear force; leptons do not.

Half-life. The time it takes for half of a given quantity of a radioactive material or a number of unstable particles to decay.

Helium. The second lightest, and second most abundant, chemical element. He^4 contains two protons and two neutrons, while He^3 contains two protons and one neutron. Atoms of helium contain two electrons outside the nucleus.

Homogeneity. A property of the universe which makes it appear the same to all observers at a given time wherever located.

Horizon problem. A quandary in standard Big Bang theory, which indicates that particles in the early universe would not have had time to be in causal contact with one another at the outset of cosmic expansion.

Hubble's law. The relation of proportionality between the velocity of separation of galaxies and their distance from each other. The Hubble Constant is the ratio of velocity to distance in this relation, and is symbolized by H or H_0.

Hubble Space Telescope. Large optical telescope in Earth orbit with a 60 inch-diameter mirror, launched in 1990.

Hydrogen. The lightest and most abundant chemical element. The nucleus of ordinary hydrogen consists of a single proton with a single electron orbiting it.

Indeterminacy principle. The postulate of quantum theory that the position and momentum of a particle cannot both be known simultaneously with perfect exactitude. Also known as Heisenberg indeterminacy or the uncertainty principle.

Inertia. The property of mass that causes it to remain at rest relative to a given reference frame, or to move at a constant velocity in a straight line in the absence of friction once in motion, unless acted upon by a force.

Inflationary theory. A theory that the expansion of the very early universe expanded much more rapidly than it does today — at an increasing rather than a constant rate.

Infrared light. Electromagnetic radiation slightly longer in wavelength than visible light.

Inverse square law. In Newtonian mechanics, the rule that a force or a measured intensity of light diminishes by the square of the distance of its source.

Ion. An atom with more or fewer electrons than normal.

Ionized. State of an atom's having fewer or more electrons than normal, leaving it with a net electrical charge.

Isotropy. The quality of being the same in all directions.

Lepton. A class of particles that carry a half unit of spin and do not participate in the strong interactions, including the electron, muon, and neutrino.

Light-year. The distance light travels in one year in a vacuum, equal to 9.45×10^{17} centimeters or about 5.8×10^{12} (six trillion) miles.

Lorentz contraction. Contraction in the length of an observed object along the direction of its motion, as seen by an external observer.

Luminosity. Intrinsic brightness of a star or galaxy.

Magnetic monopole. A massive particle with one magnetic pole.

Mass. A measure of an object's resistance to a change in the amount or direction of its motion, that is, its inertia. In the electron pair theory, mass is a manifestation of internal, localized motional energy, and is entirely due to the energy in the field of the electrons and protons it is composed of, so that there is no other "ponderable" form of matter.

Mass density, cosmic. Average amount of mass in the universe per unit volume.

Mesons. A class of strongly interacting particles, including the pi mesons, K-mesons, Rho mesons. They decay spontaneously into electrons, positrons, and neutrinos when free.

Megaparsec. One million (10^6) parsecs. Equals 3.26 million light-years.

Microwaves. Radio radiation with wavelengths of about 10^{-4} to one meter.

Milky Way. The name of the band of stars which marks the plane of our galaxy.

Muon. An unstable elementary particle of negative or positive charge, similar in its interaction with matter to the electron but 207 times heavier. Sometimes called mu- meson, but not strongly interacting like true mesons. In the electron pair theory, it consists of three charges.

Nebulae. Extended astronomical objects with a cloud-like appearance. Some nebulae are galaxies; others are actual clouds of dust and gas within our galaxy.

Neutrino. Electrically neutral particle, like a photon without rest mass, that responds to the weak nuclear force but not the strong nuclear and electromagnetic forces. It differs from a photon by having a smaller spin and much weaker absorption in matter.

Neutron star. Star with a gravitational field so intense that most of its matter has been compressed into the density of neutrons or that of an atomic nucleus.

Newton's Constant. The fundamental constant of Newton's and Einstein's theory of gravitation, symbol *G*, propotional to the strength of the attraction.

Nucleon. Proton or neutron; the constituents of atomic nuclei.

Nucleosynthesis. The fusion of nucleons to create the nuclei of atoms.

Nucleus. The central part of an atom, composed of protons and neutrons and containing nearly all of each atom's mass.

Parsecs. Astronomical unit of distance, equal to 3.2616 light-years.

Phase transition. The sharp transition of a system from one configuration to another, usually with a change in symmetry such as freezing.

Photon. Quantum of light. In the electron pair theory, a vortex ring similar to a smoke ring, always traveling at the speed of light.

Pi meson. The hadron of lowest mass. Sometimes called pions.

Planck's Constant. The fundamental constant of quantum mechanics. Symbol h. Planck's Constant was first introduced in 1900, in Planck's theory of black body radiation. It expresses the fact that the energy in a pulse of light is inversely related to its size or wavelength. When h is divided by 2π, it is the spin or angular momentum of a photon or of an electron in a circular orbit about a proton or positron.

Positron. The positively charged antiparticle of the electron.

Proton. The heavy positively charged particle found along with neutrons in ordinary atomic nuclei. The nucleus of hydrogen consists of one proton.

Quantum. Indivisible unit of energy or matter. The minimum amount of a quantity that is found in nature.

Quantum electrodynamics. Quantum theory of the electromagnetic force, which it envisions as being carried by quanta called photons.

Quantum mechanics. The fundamental physical theory developed in the 1920s as a replacement for classical mechanics and electrodynamics on the atomic and nuclear scale. In quantum mechanics waves and particles are two aspects of the same underlying entity.

Quark. Fundamental particle of large mass, usually with 1/3 or 2/3 the electronic charge from which all hadrons are made. In the electron pair theory, they are composed of very rapidly moving, massive electrons and positrons.

Quasar. Point-like source of light most frequently lying at distances of billions of light-years. Believed to be an early stage in the formation of galaxies.

Radio waves. Electromagnetic radiation with wavelengths of approximately one centimeter to a million meters.

Radioactivity. Emission of particles by unstable elements as they decay.

Radio astronomy. Study of the universe via radio wavelengths.

Radio telescopes. Radio antennae employed to detect the radio waves emitted by astronomical objects such as nebulae, active centers of galaxies, and rotating neutron stars.

Redshift. The shift of spectral lines toward longer wavelengths, caused by the Doppler Effect for a receding source. Expressed as a fractional increase in wavelength of a photon between the moment of its emission and its reception.

Relativistic. Approaching the velocity of light.

Rest energy. The energy of a particle at rest, which would be released if all the mass of the particle could be converted to energy, such as in the form of motional energy of particles. Given by Einstein's formula $E = mc^2$.

Rho meson. One of many extremely unstable hadrons. Decays into two pi mesons, with a mean life of 4.4×10^{-24} seconds.

Satellite. Object orbiting another, more massive object.

Singularity. A point of infinite mass or energy density, where the equations of general relativity and all laws of physics break down.

Special relativity. The view of space and time developed by Albert Einstein in 1905, according to which the laws of nature are independent of the constant linear motion of a system in which an observer is at rest.

Spectral lines. Bright and dark lines seen in spectra of stars and other objects emitting photons characteristic of a given atom or molecule.

Spectrum. Record of distribution of photons according to their wavelength.

Speed of light. The maximum possible velocity of photons or matter particles, equal to 299,729 kilometers per second. Symbol *c*.

Spin. A fundamental property of all particles which describes the state of internal rotation of the particle. According to quantum mechanics, the spin can have only certain values, equal to integral multiples of Planck's Constant divided by 2π for photons and mesons, half integral values for electrons, protons and neutrons.

Standard model. In cosmology, basic Big Bang theory. In quantum mechanics, the existing theories of the forces acting between nuclear particles.

Star cluster. Gravitationally bound aggregation of about 1,000 stars.

Star cluster, globular. Large, old aggregation of about a million stars.

Steady state theory. A postulate that the expanding universe was never in a state of high density, or that there was no "Big Bang." Assumes that matter is constantly being created out of empty space to keep the density of matter constant during expansion.

String theory. Theory that subatomic particles are composed of thin tubes or strings of finite diameter, and that particle properties are determined by the arrangement and vibration of the strings. Identified with vortices in the fluid ether in the present electron pair theory

Strong interaction. The strongest of the four general classes of elementary particle interaction (electromagnetic, strong, weak and gravitational).

Supercluster. A cluster of about a thousand galaxies with a typical diameter of 140 million light-years.

Supersymmetric theories. Theories that try to relate electromagnetic, nuclear and gravitational interactions.

Symmetry. A state of having a quantity that remains unchanged after a transformation such as a rotation.

Symmetry breaking. Loss of symmetry in a transformation.

Symmetry group. Mathematical group with a common property that unites its members and produces a symmetry.

Thermal equilibrium. A state in which the rates at which particles enter any given range of velocities exactly equals the rate at which they leave them.

Trillion. A thousand billion (10^{12}) .

Ultraviolet radiation. Electromagnetic waves shorter than visible light with wavelength in the range 10^{-7} cm to 2×10^{-5} cm, intermediate between visible light and X-rays.

Vacuum. Classically, empty space between particles. In quantum physics, physical space populated by virtual particles. In electron pair theory, the ether without any photons or matter particles or motion.

Virtual particles. Short-lived particles that are assumed to arise briefly as a result of fluctuations in the vacuum. In the electron pair theory, they are formed briefly as a result of the division of energetic vortex rings that divide and thus form momentary charges.

Vortex. A whirling mass of fluid that can either be open-ended, such as a tornado, or closed like a smoke-ring, with a low internal pressure, moving like a particle as a unit.

Wavelength. In any kind of wave, the distance between wave crests.

Wave-particle duality. The observed fact that photons and particles consisting of charges exhibit characteristics of both particles and waves.

White dwarf. A star that has collapsed when its nuclear fuel is used up, but not to the large density of a neutron star, or a stellar black hole.

Weak interactions. One of the four forms of elementary particle interactions, responsible for the relatively slow decays of particles like the neutron leading to radioactivity with the emission of neutrinos.

X-ray. Short-wavelength photons of high energy. X-ray energies and wavelengths lie between those of gamma rays and ultraviolet light.

ACKNOWLEDGMENTS

MANY PEOPLE OVER THE YEARS HAVE HELPED ME with their advice and encouragement, without which *Before the Big Bang* could not have come into existence. First and foremost among them is my wife Marilyn, whose love, steadfast support and wisdom sustained me in this long journey, and to whom this book is dedicated. No one could have been more fortunate in finding a friend and companion with so much courage, understanding and patience with whom to share the often lonely efforts required in the pursuit of this work.

I have been equally fortunate with my children, Daniel and Susan, a source of great encouragement over the years, and who, together with my wife, kept urging me to write a book that would allow them and their children and other non-scientists to comprehend the simplicity and beauty of the design of the universe. I was enormously lucky to have the support of my brother Arno and his wife Lila, and the large family of my wife, particularly her sister Shirley Silverman, whose first love was the books on astronomy in the library. The wonderful family I acquired when I married my wife includes her twin brother Jerry Seiner and his wife Nancy, who patiently listened to my explanations of obscure developments in particle physics and cosmology for years, and her enthusiastic sister Janice, married to Jim Colker. The warmth of this close family, together with all the many children and their husbands and wives, encouraged me to persist in my efforts.

Among those not mentioned in the book are friends whose long interest, support and advice were of enormous help in the development of the ideas and the actual preparation of the manuscript. The one I miss the most is Jens Scheer, Professor of Physics at the University of Bremen, with whom, until his recent untimely death, I shared a common interest in both the struggle to warn the public of the dangers of nuclear radiation and the desire to make physics accessible to the public. Other scientists with whom I have been fortunate to discuss the ideas in this book are the physicists Frank Meno, Allen Janis and Joe Lempert, whose diverse backgrounds and interests have helped me greatly over many years.

Among my non-scientist friends whose interest helped me greatly to undertake the task of writing this book are Shirley and Sidney Stark and Audrey and Chuck Reichblum in Pittsburgh, David Bleich in Rochester, New York, and in New York City, Lynn and David Troyka and Jay and Jane Gould. Jay was the first to read the whole manuscript and to him I owe many suggestions for clarifying aspects of the book. I am also grateful for the many years during which we shared a common concern about the environment that led to a long and fruitful collaboration, in which he enlisted his friend David Friedson.

It was through Jay that I met my editor and publisher, John Oakes, who persuaded me to finally write this book after years of procrastination. To his personal interest in this undertaking and his thoughtful and insightful editing, together with that of my wife, belong much of the credit if the basic ideas in *Before the Big Bang* are understandable to the layperson.

INDEX